AF412837

CHARACTERIZATION IN SILICON PROCESSING

CHARACTERIZATION IN SILICON PROCESSING

EDITOR

Yale Strausser

SERIES EDITORS

C. Richard Brundle and Charles A. Evans, Jr.

MOMENTUM PRESS, LLC, NEW YORK

Contents

APPLICATION OF MATERIALS CHARACTERIZATION TECHNIQUES TO SILICON EPITAXIAL GROWTH

POLYSILICON CONDUCTORS

SILICIDES

ALUMINUM- AND COPPER-BASED CONDUCTORS

TUNGSTEN-BASED CONDUCTORS

BARRIER FILMS

APPENDIX: TECHNIQUE SUMMARIES

Preface to the Reissue of the Materials Characterization Series

The 11 volumes in the Materials Characterization Series were originally published between 1993 and 1996. They were intended to be complemented by the *Encyclopedia of Materials Characterization*, which provided a description of the analytical techniques most widely referred to in the individual volumes of the series. The individual materials characterization volumes are no longer in print, so we are reissuing them under this new imprint.

The idea of approaching materials characterization from the material user's perspective rather than the analytical expert's perspective still has great value, and though there have been advances in the materials discussed in each volume, the basic issues involved in their characterization have remained largely the same. The intent with this reissue is, first, to make the original information available once more, and then to gradually update each volume, releasing the changes as they occur by on-line subscription.

C. R. Brundle and C. A. Evans, October 2009

Preface to Series

This Materials Characterization Series attempts to address the needs of the practical materials user, with an emphasis on the newer areas of surface, interface, and thin film microcharacterization. The Series is composed of the leading volume, *Encyclopedia of Materials Characterization*, and a set of about 10 subsequent volumes concentrating on characterization of individual materials classes.

In the *Encyclopedia*, 50 brief articles (each 10 to 18 pages in length) are presented in a standard format designed for ease of reader access, with straightforward technique descriptions and examples of their practical use. In addition to the articles, there are one-page summaries for every technique, introductory summaries to groupings of related techniques, a complete glossary of acronyms, and a tabular comparison of the major features of all 50 techniques.

The 10 volumes in the Series on characterization of particular materials classes include volumes on silicon processing, metals and alloys, catalytic materials, integrated circuit packaging, etc. Characterization is approached from the materials user's point of view. Thus, in general, the format is based on properties, processing steps, materials classification, etc., rather than on a technique. The emphasis of all volumes is on surfaces, interfaces, and thin films, but the emphasis varies depending on the relative importance of these areas for the materials class concerned. Appendixes in each volume reproduce the relevant one-page summaries from the *Encyclopedia* and provide longer summaries for any techniques referred to that are not covered in the *Encyclopedia*.

The concept for the Series came from discussion with Marjan Bace of Manning Publications Company. A gap exists between the way materials characterization is often presented and the needs of a large segment of the audience—the materials user, process engineer, manager, or student. In our experience, when, at the end of talks or courses on analytical techniques, a question is asked on how a particular material (or processing) characterization problem can be addressed the answer often is that the speaker is "an expert on the technique, not the materials aspects, and does not have experience with that particular situation." This Series is an attempt to bridge this gap by approaching characterization problems from the side of the materials user rather than from that of the analytical techniques expert.

We would like to thank Marjan Bace for putting forward the original concept, Shaun Wilson of Charles Evans and Associates and Yale Strausser of Surface Science Laboratories for help in further defining the Series, and the Editors of all the individual volumes for their efforts to produce practical, materials user based volumes.

C. R. Brundle   C. A. Evans, Jr.

Preface to the Reissue of
Characterization in Silicon Processing

When this volume was originally released in 1993 the chapters dealt with materials and manufacturing processes in the state of the art wafer fabrication facilities of the day. Mature fabs today still use many of those materials and processes, though state of the art fabs of today also include both different materials (e.g., Cu interconnects, organic dielectrics) and processing steps which did not exist then. After the reissue of this volume in a form close to the original, updates covering the changes that have occurred will be made and released as on-line downloads as they are completed.

C. R. Brundle and C. A. Evans, October 2009

Preface

This volume has been written to aid materials users working with silicon-based semiconductor systems. Materials problems arise in all stages of semiconductor device production: research and development of new processes, devices, or integrated circuit technologies; new process equipment definition and new process start-up; operation of state-of-the-art processes in wafer fabrication facilities; and throughout the life of each wafer fabrication process.

These materials problems are sometimes investigated using only electrical tests, but they can often be more clearly identified by using an appropriate selection of materials characterization techniques. However, the research and development scientists and engineers who work with new technologies and define or implement new processes are typically not experts in these techniques. This volume, and indeed the Materials Characterization Series, is intended to help the nonspecialist determine the best selection of techniques for a surface- or thin film materials-based problem.

This volume should be used in conjunction with the lead volume of the series, *Encyclopedia of Materials Characterization*, which defines boundary conditions for fifty widely used surface and thin-film materials characterization techniques. Each technique description discusses

- the type of information to be obtained about a sample

- appropriate samples and required sample preparation

- limitations and hardware requirements with regard to spatial resolution, compositional resolution, and sensitivity

- time required for an analysis

- destructiveness to the sample

- other important characteristics of the technique.

Each technique description also lists authoritative references for further research. The descriptions are succinct and do not discuss operation of the instruments or lengthy derivations of basic principles. They are jargon-free guidelines to aid the nonspecialist in understanding the type of information a technique provides and in selecting the appropriate technique to solve a problem.

This volume approaches materials characterization from the materials properties, processing, and problems point of view. It discusses typical materials and processes used in the manufacture of today's silicon-based semiconductor devices and provides examples of typical problems encountered in the real silicon-processing world and their identification and characterization using techniques described in the *Encyclopedia*.

The organization of the chapters in this volume is similar to the process flow of a wafer. Each material commonly used in silicon integrated circuit manufacture is the topic of a chapter, including epitaxial silicon (including silicon-germanium alloys), polycrystalline silicon, metal silicides, aluminum and copper conductors, tungsten conductors, and barrier films. Dielectric films are not covered. Each chapter discusses a typical process history of the material—deposition, thermal treatment, lithography, etc.—and the desired properties of the material, with examples of common problems seen in producing materials having the desired properties. These examples illustrate the application of appropriate characterization techniques to solve the problems.

The fifty techniques discussed in the *Encyclopedia* are the most widely used for a broad range of materials problems. Some of these techniques are seldom used in characterizing silicon-based semiconductor materials, and some techniques specific to semiconductor characterization are not included in the *Encyclopedia*. For these reasons, an appendix is provided in this volume that contains pertinent summary pages taken from the *Encyclopedia* plus lengthier descriptions of the important semi-conductor-specific methods not covered in the *Encyclopedia*.

This volume is not sufficient to make one an expert in any of the materials characterization techniques ("a little knowledge is a dangerous thing"). Its purpose is to guide one in determining which techniques to be aware of and approach first in problem-solving. Further information to help solve a materials-based problem may be obtained from the references at the close of each chapter and from experts who use characterization techniques to solve problems. (Experts are employed in the materials characterization organizations of large companies and in independent analytical service laboratories.)

I would like to acknowledge the contributions of a number of people in the preparation of this volume. Dick Brundle, the Series editor has helped beyond the call of duty in many ways. He has been patient and persistent and he has assisted in much of the editing. Gary McGuire pitched in at a time when I was unavailable and proofread all the chapters in draft form, making suggestions for improvements. Penny Strausser, my wife, was helpful in every way possible—discussing ideas, proofreading, typing—and was forgiving of my time. Finally, I thank the authors of the individual chapters for being patient and for seeing this through.

Yale Strausser

Contributors

Roc Blumenthal
Motorola, Inc.
Austin, TX

Tungsten-Based Conductors

Roger Brennan
Solecon Laboratories
Sunnyvale, CA

Spreading Resistance Analysis (SRA)

M. Lawrence A. Dass
Intel Corporation
Santa Clara, CA

Barrier Films

David Dickey
Solecon Laboratories
Sunnyvale, CA

Spreading Resistance Analysis (SRA)

C. I. Drowley
Motorola, Inc.
Mesa, AZ

Application of Materials Characterization Techniques to Silicon Epitaxial Growth

David Fanger
Intel Corporation
Rio Rancho, NM

Aluminum- and Copper-Based Conductors

N. M. Johnson
Xerox Research Center
Palo Alto, CA

Deep Level Transient Spectroscopy (DLTS)

Walter Johnson
Prometrics Corporation
Santa Clara, CA

Sheet Resistance and the Four Point Probe

David C. Joy
The University of Tennessee-Knoxville
Knoxville, TN

Electron Beam Induced Current (EBIC) Microscopy

George N. Maracas
Arizona State University
Tempe, AZ

Capacitance–Voltage (C–V) Measurements; Hall Effect Resistivity Measurements

S. P. Murarka
Rensselaer Polytechnic Institute
Troy, NY

Silicides

Philipp Niedermann
University of Geneva
Geneva

Ballistic Electron Emission Microscopy (BEEM)

Jon Orloff
University of Maryland
Washington, DC

Focused Ion Beams (FIBs)

Gregory C. Smith
Texas Instruments
Dallas, TX

Yale Strausser
Digital Instruments
Santa Barbara, CA

Roger Tonneman
Intel Corporation
Rio Rancho, NM

Chuck Yarling
Prometrics Corporation
Santa Clara, CA

Tungsten-Based Conductors

Polysilicon Conductors

Aluminum- and Copper-Based Conductors

Sheet Resistance and the Four Point Probe

Application of Materials Characterization Techniques to Silicon Epitaxial Growth

C. I. DROWLEY

Contents

1.1 Introduction

Silicon epitaxial growth has emerged as a major process technology for VLSI circuit production during the last decade. Prior to that time, silicon epitaxial growth technology had been used primarily for bipolar IC, discrete device, and power device applications. The ability to reduce latchup in CMOS circuitry by growing a lightly doped epitaxial layer (for the active device region) over a heavily doped substrate (which provides a low-resistance shunt path for substrate currents, and thus suppresses turn-on of parasitic devices) has led to the adoption of epitaxy for high-volume CMOS processes.[1] Silicon epitaxial growth is also a critical process for the production of high-performance circuits incorporating both bipolar and CMOS devices (i.e., BiCMOS technology).[2]

Epitaxial growth applications have expanded to include the selective epitaxial growth of silicon on patterned substrates. Selective epitaxy has been demonstrated on a number of VLSI structures. Some examples of such selective growth applications are the creation of low-encroachment isolation,[3, 4] elevated MOS source/drain formation,[5] and DRAM cells.[6, 7] Three-dimensional structures such as folded CMOS inverters have also been fabricated.[8]

Epitaxial growth of $Si_{1-x}Ge_x$ alloys on silicon has attracted considerable interest because of the smaller bandgap of the alloy films. The ability to perform bandgap engineering in a silicon-based alloy system allows a number of exciting device applications previously confined to III–V materials systems. Very high-speed heterojunction bipolar transistors (HBTs)[9] using $Si_{1-x}Ge_x$ alloy bases have been demonstrated. Heterojunction bipolar transistors also show advantages over conventional silicon homojunction bipolar transistors for low-temperature BiCMOS operation.[10] Modulation doping also has been demonstrated in the $Si/Si_{1-x}Ge_x$ system.[11] The smaller bandgap of the $Si_{1-x}Ge_x$ alloys also allows formation of detectors (on a silicon substrate) useful at the wavelength of modern fiber-optic transmission systems.[12] The bandgap difference between the alloy film and silicon also may be exploited for optical-waveguide applications.[12]

Silicon-based epitaxial films serve a variety of functions in device manufacture. Most commonly, the films provide a region for active device fabrication (e.g., in BiCMOS). Epitaxial films also may serve as key device elements (e.g., the epitaxial base of an HBT).

Because of the complex interrelationship between the epitaxial films and the final device properties, a number of material parameters are of critical importance for successful device fabrication. These include (among others) crystal quality, film resistivity and thickness (and their variation over the growth surface), dopant profiles, and alloy composition (for $Si_{1-x}Ge_x$ films). Such material parameters are affected by the growth process and by pretreatments such as in situ precleans. As we will see, the increased sophistication of epitaxial growth processes combined with the expanding number of critical material parameters has led to increased dependence on sophisticated analytical techniques for process characterization.

This chapter examines the conventional epitaxial growth of silicon on silicon substrates and then covers selective silicon growth and $Si_{1-x}Ge_x$ heteroepitaxial growth on silicon. Each of the three topics is introduced by a brief review of the growth technology and concepts, followed by a discussion of the characterization techniques appropriate to the material produced by each method. These techniques include sophisticated analytical techniques and methods suited to routine use in a manufacturing environment. Most of the characterization techniques are discussed in the lead volume of this series, *Encyclopedia of Materials Characterization*; consequently, emphasis is on illustrating applications and limitations of the techniques.

1.2 Silicon Epitaxial Growth

Basic Chemical Reactions

Silicon epitaxial growth by chemical vapor deposition can employ a number of reactions. Most commonly, a reactant gas such as a chlorosilane is diluted in a carrier gas such as hydrogen and passed over a heated silicon substrate. Epitaxial growth occurs by the surface reaction of the silicon source gas at the elevated temperature to produce silicon, hydrogen, or HCl. Common silicon source gases and their net reactions are

$$SiCl_4 + 2H_2 = Si + 4HCl$$
$$SiHCl_3 + H_2 = Si + 3HCl$$
$$SiH_2Cl_2 = Si + 2HCl$$
$$SiH_4 = Si + 2H_2$$
$$Si_2H_6 = 2Si + 3H_2$$

The choice of reactant is dictated both by the particular application and by economic considerations. $SiCl_4$ and $SiHCl_3$ are comparatively inexpensive sources of silicon (by cost per mole); $SiHCl_3$ is widely used in cost-sensitive applications (e.g., epitaxy for CMOS). These sources require comparatively high temperatures and are replaced by SiH_2Cl_2 for lower-temperature applications where dopant profile control is important. The silicon hydrides are used in low-temperature applications.[13, 14] SiH_4 also is used in specialized applications where the presence of chlorine is undesirable (e.g., silicon-on-sapphire). Both SiH_4 and Si_2H_6 are useful in growth on patterned substrates when epitaxial growth is desired (on single-crystal material exposed through windows in the masking material) at the same time as polysilicon growth on the masking material.

Precleaning Considerations

Growth of a defect-free epitaxial film requires an initial silicon surface free of damage, contaminants, or masking films such as silicon dioxide. Native oxides readily form on silicon, so that conventional epitaxial growth methods provide some technique for in situ precleaning of the silicon surface. Such precleans historically have included high-temperature (1100 °C) surface etches using HCl, or high-temperature bakes (typically in an H_2 ambient). The HCl etch process removes the surface silicon to some depth, whereas the high-temperature bake process allows silicon dioxide reduction according to the reaction

$$Si + SiO_2 = 2SiO(gas)$$

The temperature at which this reaction is effective is dependent on the partial pressure of oxidizers (e.g., O_2 or H_2O) in the system.[15, 16] The presence of hydrogen allows lower-temperature oxide reduction than in a vacuum alone for a given oxidizer partial pressure.[17, 18]

Several alternative precleaning techniques are discussed in "Preclean Quality" in Section 1.3.

Reactor Types

Several different commercial reactor designs are available for epitaxial film production. One common type is the radiantly heated "barrel" configuration (Applied

Materials, Inc., Santa Clara, CA), in which wafers rest vertically in pockets on a prism-shaped susceptor in a cylindrical chamber. Lamp heating is used, and the susceptor rotates during deposition. Reactant gases are introduced at the top of the chamber and exhausted at the bottom. A second type is the "vertical" or "pancake" reactor (Lam Research, Inc., Fremont, CA), in which wafers rest on a radio frequency (RF) induction-heated horizontal annular susceptor. Reactant gases are injected vertically through the center of the susceptor and pass over the wafers during recirculation through the chamber prior to exhausting at the bottom of the chamber.

The use of larger silicon wafer diameters has limited the productivity of the barrel and vertical reactor designs. Two divergent approaches are being used to improve productivity. The first approach is a "radial" design (Questor Technology, Inc., Fremont, CA), in which wafers are placed on both sides of vertically mounted susceptor segments arranged radially on a large carrier. Gases are injected from outside and are exhausted through a center port. This design handles up to fifty 200-mm-diameter wafers in one load. The second approach is a single-wafer, horizontal reactor with a radiantly heated susceptor (ASM Epitaxy, Inc., Tempe, AZ). The single-wafer reactor throughput is optimized by using very high growth rates and in situ cleaning of the chamber during loading and unloading of wafers. This design is comparatively low-cost and can exceed the productivity of the "barrel" and "vertical" reactor designs for epitaxial growth on 200-mm-diameter wafers.

In addition to these commercial reactors, several experimental reactors have been devised for low-temperature applications. Such reactors are described in Section 1.5, "$Si_{1-x}Ge_x$ Epitaxial Growth."

1.3 Film and Process Characterization

Critical parameters for epitaxial films include the crystal quality (including surface roughness, particulate contamination, extended defects such as dislocations and stacking faults, point defects, and deep-level impurities), thickness, resistivity or dopant concentration, and dopant profiles in the films. Each of these subjects are considered in turn.

Crystal Quality

One of the first parameters characterized in an epitaxial growth process is the crystal quality. Advanced high-density integrated circuit process requirements dictate production of high-quality epitaxial films with epitaxial defect densities $\ll 1$ cm^{-2}.

Optical microscopy enhanced by phase-contrast techniques (e.g., Nomarski interference contrast) often is used to examine epitaxial films. ASTM standard F 522-88[19] describes the use of the interference-contrast technique to test for grown-in stacking faults. The technique sensitivity is dependent on the film thickness and the area scanned. The surface area of the trace of a growth stacking fault pyramid on (100) silicon is approximately 2 × (film thickness)^2. Consequently, a magnification of ~50× is needed to resolve features on stacking fault traces in a 5-μm-thick film, whereas a magnification of ~250× is required for the same fault-trace resolution in a 1-μm-thick film. Because the field of view at such magnifications is limited, the distance scanned to determine defect densities in the 1-cm^{-2} range is quite large. Typical inspection patterns include one or more scans across the wafer diameter.[20]

The multiple-scan optical microscopy method for defect inspection discussed above is time-consuming. However, optical measurement of surface quality and defect density can be automated by using laser scatterometry.[21] Commercial systems can rapidly scan the whole surface of a silicon wafer to determine the location of light-scattering defects. Defect size is estimated from the amount of light scattered by the defect. System calibration is performed by scattering light from spherical objects of known size and may be inaccurate for defects with strong crystallographic orientations (e.g., epitaxial spikes) or minimal surface relief (e.g., stacking faults). Accurate identification of defects typically requires additional microscopic examination. Automated systems which use defect coordinates from the scattering measurement to control microscope positioning have been developed[21] to allow rapid inspection of each detected defect. Laser-scatterometry techniques, used routinely in defect-reduction efforts, are capable of detecting defects with an effective diameter ≥0.3 μm. Higher-resolution equipment is in development.

Some defects (such as dislocations) are not detected readily by optical techniques. Surface-sensitive etches which preferentially attack defects[22–24] may be used to reveal dislocations, stacking faults, and other surface defects in more detail. The defect etch approach may be used in combination with either optical microscopy or scanning electron microscopy (SEM) to determine defect type and density. Optical examination is best suited for comparatively thick films (typically >2 μm thick), since ~1 μm of silicon must be removed during the etch to generate a feature detectible in optical microscopes.[25] Thinner films can be etched for shorter times and examined at higher magnification using an SEM. Examples of etched defects are shown in References 22–25.

Crystal-quality characterization of thinner films also can be accomplished by other techniques. Evidence of epitaxial orientation can be determined nondestructively using Rutherford backscattering spectrometry[13] (RBS) or SEM electron channeling patterns.[3] These methods are relatively insensitive to the presence of defects (>10^6 cm^{-2}) and are useful primarily for screening. Somewhat higher resolution defect density estimates can be obtained from cross-sectional transmission electron microscopy (XTEM) or plan-view TEM (PTEM). The size of the sampled region (up to ~0.1 μm × 500 μm for XTEM, and up to ~(300 μm)^2 for PTEM) limits defect density sensitivity to ~10^6 cm^{-2} for XTEM and ~10^3 to 10^4 cm^{-2} for PTEM.

Measurement of low defect densities, below the sensitivity limits of the physical techniques mentioned above, may be performed with a number of electrical techniques. These include MOS capacitance-time measurements to extract generation lifetimes, junction leakage measurements in diodes and bipolar transistors, electron-beam-induced current (EBIC) measurements in diodes or bipolar transistors, emitter-collector leakage current (I_{ceo}) measurements to look for "pipe" shorts in bipolar transistors, junction breakdown characteristics, and deep-level transient spectroscopy (DLTS). Some applications of these techniques are given in "Preclean Quality" (following), "Defect Density and Growth Morphology" elsewhere in Section 1.3, and "Bandgap Measurements" and "Interfacial Abruptness and Outdiffusion" in Section 1.6; References 26 and 27 provide details on electrical characterization techniques and their applications.

Preclean Quality

A key factor in defect-free epitaxial film production is the wafer surface cleanliness prior to growth. Several in situ and ex situ preclean processes have been studied. Although the early emphasis of preclean processes was defect-free film production, recent preclean processes have added constraints of low-temperature operation and minimal surface material removal in order to preserve dopant profiles already in the silicon substrate.

Historically, precleaning effectiveness has been verified (indirectly) after epitaxial growth using the defect-detection techniques listed in the previous section. A variety of modern material analysis techniques have been applied to direct studies of the precleaning process prior to growth. Information regarding the surface structure, adsorbed species, surface bonding, and the effects of chemical and thermal processes on the substrate surface have been obtained using (among others) Auger spectroscopy, reflected high-energy electron diffraction (RHEED), ellipsometry, thermal desorption spectroscopy, and internal-reflection infrared (IR) spectroscopy. Such studies have provided valuable insights both into the mechanisms of standard precleaning processes and into new precleaning methods for advanced applications, as will be seen below.

HCl etching This technique involves exposure of the silicon surface to HCl gas at an elevated temperature (typically >1100 °C) so that surface etching occurs. This process is extremely effective in removing residual mechanical damage from polishing.

Chlorine-containing gases such as HCl and the chlorosilanes may react with metals (e.g, in the source container or gas plumbing) in the presence of small amounts of water. These metals then may be carried along with the reactant gas and incorporated in the epitaxial film. DLTS has been used to quantify Fe, Cr (as CrB), and Ti concentrations in epitaxial films as a function of preclean process and silicon source gas[28] The use of $SiCl_4$ source gas after a 5-μm HCl etch resulted in concentrations of 0.5–1 × 10^{12} cm^{-3} [Fe], 0.6–1 × 10^{12} cm^{-3} [CrB], and 5 × 10^{11} cm^{-3} [Ti], respectively. SiH_2Cl_2 source gas yielded substantially lower levels of metals (0.5–1 × 10^{11} cm^{-3} [Fe], 2 × 10^{11} cm^{-3} [CrB], and 4 × 10^{10} cm^{-3} [Ti]) than $SiCl_4$ in this study.[28]

High-temperature prebake The HCl etch process may remove a significant thickness (from 0.1 μm to >1 μm) of the original substrate. Such etching can alter diffused regions in the substrate. For example, buried n^+ subcollector diffusions

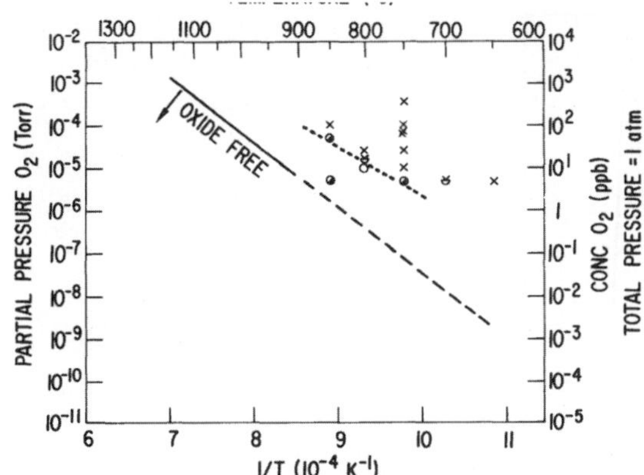

Figure 1.1 Arrhenius plot of the boundary between oxide-free and oxidized silicon. The solid/dashed line shows the boundary under UHV conditions;[15, 16] the data and dotted line show the boundary in the presence of 1 atm H_2[17, 18]. Crosses indicate surface oxide found; circles indicate oxide-free surface. (After P. D. Agnello and T. O. Sedgwick, IBM T. J. Watson Research Center[17, 18]; reprinted by permission of the authors.)

used in modern bipolar and BiCMOS technologies may be ~1 μm deep, and tight control of the resistivity of these regions is necessary. Etching of the substrate during precleaning can remove a substantial fraction of such a diffusion and greatly increase the diffusion sheet resistance. Hence, such etching cannot be tolerated in these technologies. Consequently, precleaning practice today commonly uses a high-temperature bake in hydrogen, rather than an etch, to clean the surface. The effect of high-temperature bakes on surface cleanliness has been the subject of a number of studies. Auger spectroscopy, ellipsometry, and RHEED were used to study the removal of oxygen- and carbon-containing species in a special reactor design which allowed transfer from a growth chamber to an analysis chamber. Temperatures of about 800 °C were required to desorb physisorbed species, whereas temperatures greater than 900 °C were required to reduce the oxide coverage.[29]

The reactions of O_2 and H_2O with Si have been determined as a function of temperature and pressure in a UHV system. Optical and scanning electron microscopy were used to examine the Si surface after processing, and the equilibrium boundary curves between regions of stable SiO_2-covered Si and clean Si were established[15, 16] (Figure 1.1). This finding led to experimental reactors[13] with very low background pressures of oxidizers, which allowed precleaning at temperatures ≤800 °C while maintaining reasonable epitaxial quality.

Chemical cleaning processes which leave a thin oxide on the Si surface are commonly used prior to epitaxial growth. The desorption of such oxides was studied (again, under UHV conditions) using a combination of Auger spectroscopy, RHEED, and thermal desorption spectroscopy.[30] SiO was determined to be the

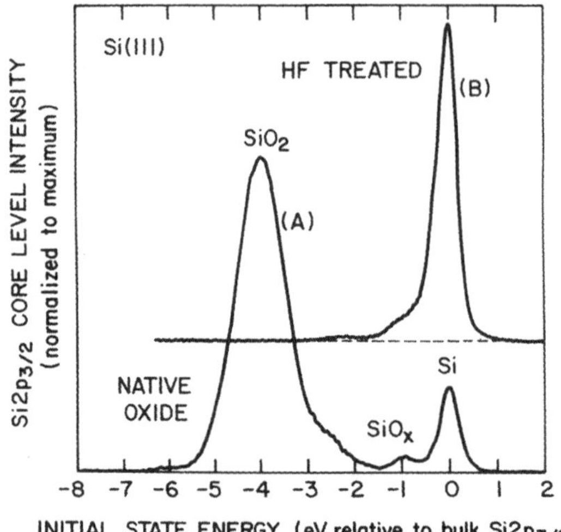

Figure 1.2   Si *2p* photoelectron spectra of (111) Si surface showing (A) the presence of oxide prior to HF cleaning and (B) the oxide-free surface after 10-min exposure to room air following the HF clean. (After Reference 33; reprinted by permission of the authors.)

primary desorption product. The desorption temperature was a function of the specific chemical cleaning process; the observed variation was explained by both the variation in oxide thickness grown by the different processes and by differences in the interfacial structure between the different oxides and the substrate. Evidence for inhomogeneous desorption through void formation in the oxide films was obtained using RHEED.

The above bake studies emphasized UHV conditions, although H_2 is usually present during in situ epitaxial precleaning processes. The stability of SiO_2 in the presence of one atmosphere of H_2 was studied for various oxidizer partial pressures using secondary ion mass spectrometry (SIMS) to examine oxygen concentrations at the epi/substrate interface, together with surface SEM examination for defects indicating incomplete oxide removal prior to growth.[17, 18] The presence of hydrogen decreases the stability of the SiO_2 film, so that higher oxidizer pressures can be tolerated while still maintaining an oxide-free surface (Figure 1.1).

Ex situ cleaning   The silicon surface exhibits increased resistance to oxidation after treatment in aqueous HF solutions. The surface of HF-treated (111) silicon has been studied using internal-reflection IR spectroscopy and found to be primarily hydrogen terminated with a mix of mono-, di-, and trihydrides.[31] X-ray photo-electron spectroscopy (XPS) studies of HF-treated (100) surfaces have detected retarded oxidation rates in room air.[32] Increased oxidation was correlated with the removal of the hydrogen during thermally stimulated desorption.

Aqueous HF treatment has been used successfully as the preclean prior to high-quality epitaxial growth in the 425–650 °C range.[33] Photoelectron spectroscopy

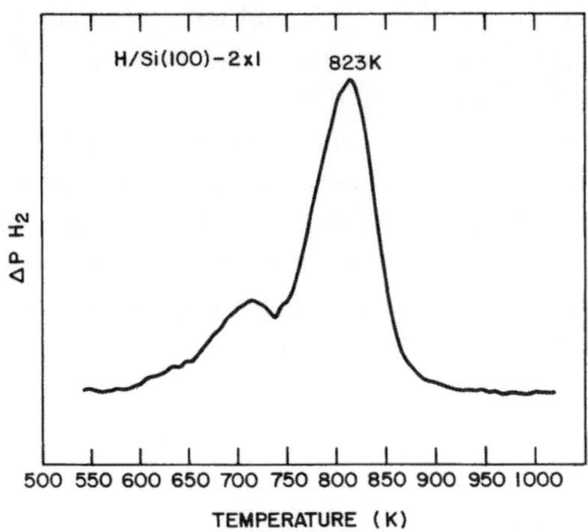

Figure 1.3 Thermal desorption spectrum of an HF-cleaned hydrogen-passivated silicon surface showing hydrogen desorption above ~625 °C. (After Reference 33; reprinted by permission of the authors.)

has confirmed that the surface remains oxide-free after HF treatment and 10 min exposure to room air (Figure 1.2), and thermal desorption spectroscopy has shown that hydrogen evolution from the silicon surface occurs above ~625 °C (Figure 1.3). In the same study, epitaxial growth was noted to be difficult to achieve from ~650 °C up to 750 °C, since oxide could form on the surface due to incomplete passivation. Above 750 °C the normal thermal reduction of SiO_2 occurred, allowing defect-free growth.

Plasma preclean Hydrogen plasmas provide another means of hydrogen passivation of the silicon surface.[34] Auger spectroscopy showed that hydrogen plasma treatment removes oxygen and carbon from the surface, and RHEED patterns have been used to infer the presence of regular mono- and dihydride termination. Subsequent Auger examination showed that the plasma-induced passivation retarded surface oxidation in air. Fourier-transform infrared spectroscopy (FTIR) and XPS, used in combination, have shown that initial oxidation of the hydrogen-plasma-passivated surface proceeds through attack of Si–Si bonds, rather than Si–H bonds.[35]

Argon sputtering also has been employed for precleaning. Optimization of the Ar sputter-clean process involved TEM interfacial studies and SIMS analysis of residual oxygen and carbon at the interface. Optimized precleaning allowed fabrication of bipolar transistors with highly ideal I–V characteristics in epitaxial films grown at 800 °C.[14]

Thickness

High-performance bipolar and BiCMOS circuits require tight epitaxial layer thickness control in order to minimize variations in bipolar device parameters such as transit frequency, breakdown voltage, and collector-base capacitance, and CMOS device parameters such as junction capacitance and the threshold-voltage sensitivity to substrate bias (body effect). Epitaxial thickness measurement techniques include: IR reflectance, spreading resistance, bevel and stain, and SIMS.

Infrared reflectance The most common thickness measurement method uses the wavelength variation of IR light reflectance from epitaxial layers grown on heavily doped regions.[36] As the wavelength is varied, the reflectance exhibits a pattern of minima and maxima because of interference between light reflected by the heavily doped region and light reflected at the epitaxial film surface. The minima or maxima wavelengths are related to the epitaxial thickness by

$$\lambda_{min/max} = \frac{4\pi T(n^2 - \sin^2\theta)^{1/2}}{j\pi - \phi_1 + \phi_2}$$

where T is the epitaxial film thickness, n is the silicon refractive index, θ is the angle of incidence, ϕ_1 and ϕ_2 are the phase shifts at the air/epi and epi/heavily doped region boundaries, respectively, and the order index j is odd for minima and even for maxima.

The accuracy of this method decreases with a decrease in the number of maxima and minima for the wavelength range covered (typically wavenumbers $\approx$ 400–4000 cm^{-1}, or wavelengths from ~2.5 to 25 μm). The technique is limited to film thicknesses >0.5 μm in order to obtain at least one full cycle (two minima, one maxima) in the normal wavelength range. This technique is repeatable on a given instrument to within 2% (one standard deviation) for epitaxial layers >2.5 μm thick.[36] For thinner layers, cross-calibration against other techniques (e.g., spreading resistance or SIMS) is needed for best precision (Figure 1.4).

The interference signal depends on the reflectance at the epi/heavily doped region interface. This reflectance is determined by the carrier concentration in the heavily doped region. For a sufficient interference signal to be ensured, it is recommended that the epitaxial film resistivity is >0.1 Ω-cm at 23 °C and the resistivity of the heavily doped region under the epi is <0.02 Ω-cm at 23 °C.[36]

The FTIR spectroscopic technique[37] is commonly used for this measurement, since the Fourier transform method has the advantages of being fast (a few minutes per sample), nondestructive, and well-suited to automation for routine process monitoring.

Spreading resistance profiling Spreading resistance profiling (SRP) can be used as an alternative to FTIR for thickness measurement as long as the epitaxial film resistivity (or type) differs from the substrate. Depth resolution is limited (by factors noted below), and the measurement of extremely thin epitaxial layers (<0.2 μm thick) is accomplished more accurately by elemental profiling (e.g., using SIMS). Deposition of an oxide on the epitaxial surface prior to beveling the sample improves the depth accuracy of the technique, since any rounding of the bevel edge can be limited to the oxide layer.[38] The oxide also provides a very high resistivity region for accurate determination of the silicon surface position. A destructive

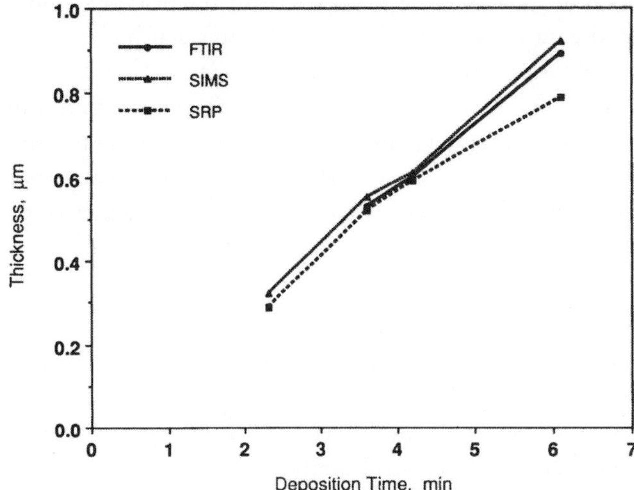

Figure 1.4 Comparison of epitaxial thickness measured by SIMS, FTIR, and SRP on samples grown for different deposition times. Using this cross-calibration, one can take into account the slight offset between SIMS and FTIR when measuring submicron epitaxial layers. (Courtesy L. K. Garling, Motorola, Inc.)

analysis method, SRP is suited to applications where IR reflectance is ineffective (e.g., in the absence of a heavily doped buried layer, or in the case of a heavy doping concentration in the epi).

The SRP epitaxial thickness measurement accuracy can be affected by depletion regions or dopant-profile gradients.[39] Since the SRP technique depends on the carrier concentration, the measured *electrical* junction depths may be either shallower or deeper than the metallurgical junction, depending on surface charge and on carrier "spillage" from the actual dopant profile. SRP accuracy may be increased by using data-reduction techniques that solve the Poisson-Boltzmann equation.[39, 40] Accuracy can be improved further by starting the data reduction with a close estimation of the dopant profile (e.g., from simulation or SIMS).[40]

Bevel-and-stain   This technique relies on changes in carrier concentration or type between different layers to allow differential chemical staining. Sample preparation involves either angle lapping[41] or grooving[42] of the sample surface, followed by exposure to a staining solution. A variety of staining solutions have been formulated[41]; the choice of solution depends on the junction type. The stain location depends on the carrier concentration at the bevel surface and is subject to errors caused by carrier "spillage" and surface charge,[39] much like SRP.

Secondary ion mass spectrometry   SIMS also may be applied to epitaxial layer thickness determination. SIMS thickness measurements depend on the correlation between the dopant concentration and the layer thickness. SIMS is applicable only when dopant concentrations are high enough to be detected (i.e., greater than about 10^{15} to 10^{17} cm^{-3}, depending on dopant species and instrument type). Thickness is then determined from the measured sputter time and crater depth. This technique is extremely useful in evaluating thin (<0.5 μm) epitaxial layers[43] and readily resolves structures with multiple thin layers.[44] SIMS also provides a good calibration reference for FTIR measurements, especially for the measurement of submicron film thicknesses. A destructive technique, SIMS is comparatively time-consuming (particularly as film thicknesses exceed 1–2 μm) and is most appropriate in cases where the other methods fail.

Dopant Concentration and Dopant Profiling

Dopant concentration in epitaxial films is measured by a variety of techniques which give either the carrier concentration or resistivity (four-point probe, SRP, capacitance–voltage, Hall effect), or elemental concentration (SIMS). Dopant profiles may be obtained using SRP, C–V, Hall sectioning, or SIMS methods.

Four-point probe   Epitaxial layer sheet resistivity usually is measured by the four-point-probe technique. Sheet resistivity data are combined with thickness data (obtained using the methods in "Thickness" elsewhere in Section 1.3) to obtain the epitaxial resistivity. This technique is suitable for measuring epitaxial layer resistivity on opposite-type substrates (i.e., *n*-epi on *p* substrate or vice versa) so that inclusion of the substrate conductivity is avoided. If the epitaxial layer is the same type as the product substrate, an opposite-type substrate commonly is added to the growth run specifically for resistivity measurements. The depletion layer (especially of high-resistivity epitaxial layers) at the epi/substrate junction must be taken into account when calculating the resistivity.

The precision of this technique decreases as the resistivity range of the epitaxial film increases.[45] The technique leaves mechanical damage from the probes on the tested sample. Probe penetration limits the usefulness of this technique on very thin layers (~0.1 μm).

SRP   Spreading resistance provides epitaxial-layer carrier concentration information. SRP can measure a wide concentration range (~10^{14} to 10^{20} cm^{-3}). SRP measurement accuracy is dependent on careful probe and sample preparation together with calibration on standard samples.[38] The resistivity obtained by this technique has a precision of about 20%.

In addition to carrier concentration measurements in the grown film, SRP is used to examine carrier profiles caused by deliberate changes in concentration (e.g., from a heavily doped region into a lightly doped epitaxial layer) and by unintentional effects (e.g., autodoping). One must take into account the uncertainty in carrier and dopant profiles caused both by carrier spillage and by the measurement technique in order to obtain accurate profiles.[39, 40]

The wide carrier-concentration sensitivity range makes SRP ideally suited to studies of epitaxial autodoping effects. Epitaxial films grown over either heavily doped substrates or substrates with patterned heavily doped surface regions may incorporate dopant from the heavily doped region during growth. Such incorporation occurs by dopant evaporation from the heavily doped region followed by readsorption on the growth surface and incorporation into the growing film. Autodoping phenomena can have significant electrical effects. For example, in technologies with a patterned arsenic-doped n^+ buried layer (e.g, some BiCMOS processes), uncontrolled arsenic autodoping can result in undesired n-type doping of the epi over regions between buried layers ("lateral autodoping"), shorting adjacent buried layers. Spreading resistance profiles of epitaxial regions that are not over the heavily doped buried layers are used routinely during process optimization to minimize such unwanted "lateral" autodoping. SRP has been used to verify models for arsenic dopant incorporation and autodoping phenomena.[46]

Capacitance–voltage Capacitance–voltage (C–V) techniques can be used to determine both the carrier concentration and concentration profile in epitaxial films. A variety of structures can be used for C–V measurements, including p–n junctions, MOS capacitors, and Schottky barriers.

Schottky barriers are particularly attractive because they can be formed with minimal processing. Formation methods include evaporation of metallic layers onto the surface to form Schottky barriers, or sintering of metal films to produce metal–silicide barriers. Another common rapid formation technique uses liquid mercury as the barrier contact material.[47] Measurements using Schottky barriers are useful for characterizing the carrier concentration of epitaxial layers grown on the same-type substrate. Since the depletion-layer capacitance can be measured as a function of voltage, information on carrier profiles and layer thickness can be obtained.

The C–V technique has a number of limitations. Layer thickness measurements are limited by breakdown voltage (for thick or heavily doped layers).[47] Epitaxial layers must also be thicker than the zero-bias depletion width. The carrier concentration is "averaged" over the Debye length, so that abrupt dopant concentration changes will appear to be broadened.[48] Surface preparation is important for reproducibility with Schottky barriers, as is accurate knowledge of the barrier area and parasitic capacitance from the measurement apparatus.[47]

Hall effect The Hall effect may be used to characterize the mobility and carrier concentration[49] in epitaxial films. This technique is more complicated than previous methods since specialized measurement structures are required. Hall effect measurements may be combined with anodic sectioning to provide depth characterization of carrier concentration and mobility.[50] This depth-characterization method is useful primarily for thin (<0.5 μm) films.

SIMS SIMS is very useful when elemental information is desired, or when thin films are being examined. The SIMS technique can detect minimum elemental concentrations of dopants in the range $\sim 10^{15}$ to 10^{17} cm^{-3}, depending on the dopant species and instrument type and mode of operation. The accuracy of the measured concentration depends on calibration against a standard.

SIMS is useful in situations where more than one dopant may be present (e.g., autodoping). SIMS is particularly effective for characterizing thin layers (<0.5 μm) and abrupt profiles. Careful characterization of the profile as a function of beam

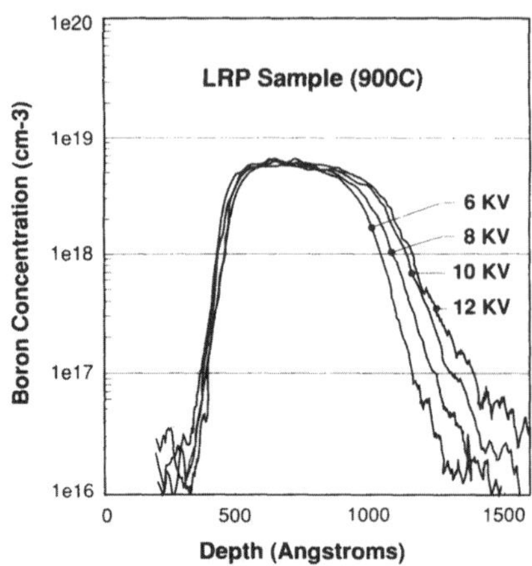

Figure 1.5   SIMS profiles of LRP-grown boron-doped Si epitaxial layer. The primary beam energy has been varied; by extrapolation of the profile slopes to 0 kV, the profile abruptness is estimated to be <50 Å/decade. (After Reference 43; reprinted by permission of J. E. Turner, Hewlett-Packard Company.)

voltage has allowed measurement of boron concentration transitions as steep as 50 Å/decade (Figure 1.5).[43]

1.4 Selective Growth

Basic Process Considerations

The selective silicon growth process is shown schematically in Figure 1.6. A masking material (typically SiO_2 or Si_3N_4) is patterned to expose regions of the silicon substrate; these exposed windows can vary in size from sub-micrometer to millimeter dimensions. The masked substrate then is exposed to reactant gases. Silicon nucleation on the masking material must be suppressed for growth to occur selectively on the exposed substrate regions. Selective growth commonly is performed using a chlorosilane reactant gas. Nucleation suppression may be achieved by adding an etchant gas such as HCl to reduce the Si supersaturation in the reactant mixture,[3, 4, 51] although it is not necessary to add HCl if the deposition pressure, temperature, and Si/Cl and Cl/H ratios are adjusted so that Si supersaturation is low (<10%).[52] Nucleation also can be suppressed by alternating growth and etch cycles.[53] In this latter alternative, the growth cycle is kept short compared to the finite nucleation time of polycrystalline silicon on the insulator, so that the etchcycle completely removes incipient nuclei without completely removing the epitaxial layer grown during the growth cycle.

Selective growth typically is performed at lower temperatures (800 to 950 °C) and pressures (10 to 80 torr) than conventional epitaxial growth. Lower growth

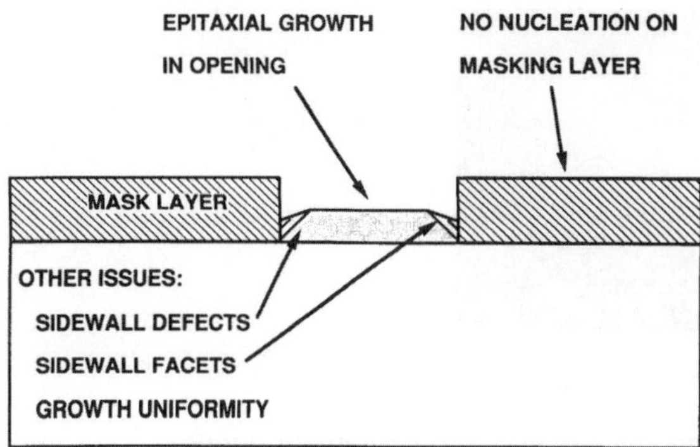

Figure 1.6   Schematic representation of the selective growth process.

temperatures favor improved uniformity and lower defect density; lower growth pressures enhance selectivity.

The presence of the patterned surface presents material and process challenges not found with conventional unpatterned epitaxial layers. Critical parameters unique to selective growth include selectivity, crystal quality and planarity adjacent to the masking material, masking material integrity, and thickness uniformity (both within a window opening and across a patterned surface).

Defect Density and Growth Morphology

The defect characterization techniques discussed in "Crystal Quality" in Section 1.3 can be used for selective epitaxial material. However, small growth-window sizes and the relatively thin layers typically grown (often <1 μm thick) frequently preclude the use of optical inspection for epitaxial defect studies. Detailed defect studies have relied on electron microscopy (either SEM or TEM) and electrical techniques. Selectivity typically has been determined using optical microscopy of the mask regions.

SEM   SEM examination allows estimation of defect density, especially around the edges of small seed windows. When high tilt angles are used, surface roughness related to polycrystalline regions or other defects can be detected readily.[4, 54, 55] The high tilt angle technique has been used in studies of defect density as a function of pressure and temperature.[54] SEM examination after defect decoration etching also has been employed to study film quality.

SEM also has been used to study growth morphology.[3, 4, 56] Selective silicon films can form facets adjacent to the masking material. The facet orientation depends on the crystallographic orientation of the material adjacent to the sidewall, and the growth conditions. Facet-free material has been obtained in (100) epitaxial material for growth along sidewalls parallel to {100} planes, whereas facets tend to form along {110} sidewalls[3, 4] (Figure 1.7). These facets can be {111}, {311}, or higher planes, depending on process conditions.[3, 4, 56] Facet growth rates along {110} sidewalls are increased by the presence of the oxide.[57]

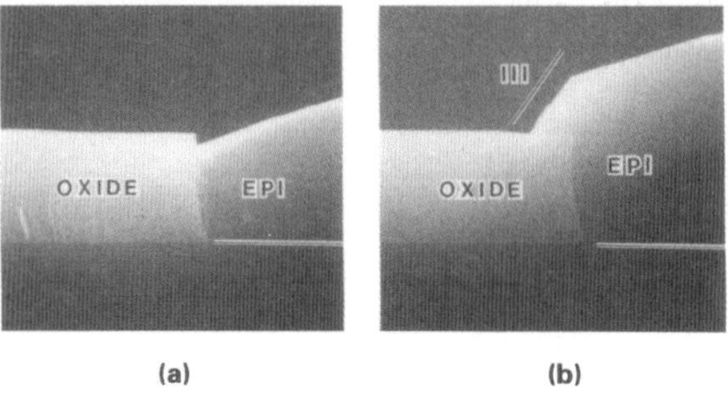

(a)   (b)

Figure 1.7   Cross section SEM micrograph showing facet evolution during (100) silicon selective epitaxy along a {110} oxide sidewall. A {311}-type facet is seen prior to overgrowth; after overgrowth, a {111}-type facet appears. (From Reference 57; reprinted by permission of the authors.)

TEM   TEM reveals more detail about selective epitaxial film defects. PTEM of (100) epitaxial films grown along {100} sidewalls has shown a much smaller defect density than in films adjacent to {110} sidewalls.[4] Plan-view studies have shown that the selective epitaxial film defect density on (111) substrates also depends on the mask sidewall orientation.[4] XTEM has been used to study defects in (100) epitaxial material along oxide sidewalls oriented parallel to {110} planes; diffraction patterns and HRTEM images demonstrated that the defects were twins[57] (Figure 1.8).

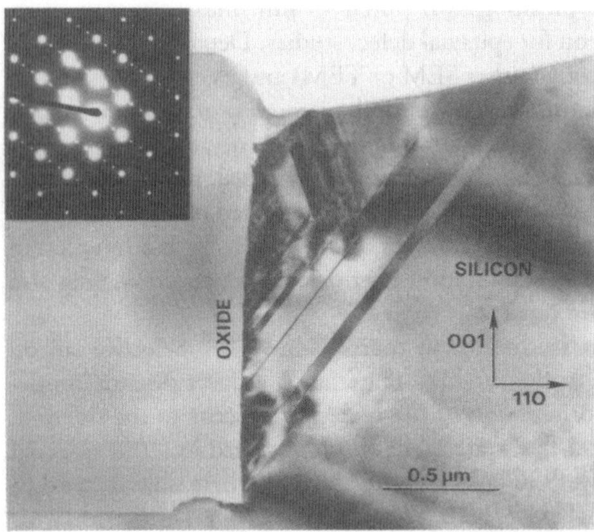

Figure 1.8   XTEM micrograph of a (100) selective epitaxial film at a {110} oxide sidewall. The laminae are twins, as shown by the diffraction pattern. (From Reference 57; reprinted by permission of the authors.)

TEM study also has revealed that defect density decreases as the growth temperature decreases.[4, 55]

Electrical techniques Electrical techniques have been used to characterize both the epitaxial defect density and the interfacial properties between the epitaxial film and the masking material.

Reverse-bias leakage in n^+–p diodes has been used to compare the electrical defect density in selective epitaxial films grown against {100} and {110} sidewalls, with either SiO_2 or Si_3N_4 as a masking material.[58] Devices built using a standard locally oxidized silicon (LOCOS) isolation were used as a control. The leakage was highly perimeter-dependent, indicating that the leakage originated at the sidewalls. The {100} oxide sidewall resulted in lower leakage than the {110} sidewall, in accordance with the observed defect behavior discussed earlier in this section.

Diode leakage reveals other geometric effects in selective epitaxial growth. Although sidewall defects can be minimized by sidewall orientation along ⟨100⟩ directions, facets and defects can still form in the mask window corners. The reverse leakage of diodes produced with differing numbers of corners is strongly dependent on the number of corners (Figure 1.9). A reduction in growth temperature significantly reduces the leakage related to the corner regions. Similar reductions in leakage with decreases in growth temperature also have been seen in sidewall diodes.[56]

Generation lifetimes in selective epitaxial films have been studied using MOS capacitor structures,[59] and also using a novel sidewall-gated diode structure.[60] Lifetimes near the sidewall were estimated to be ~10 ns in both studies, which was significantly lower than the 15–100-μs values obtained away from the sidewalls. Recombination lifetimes of ~200 μs in regions away from the sidewalls have been inferred from bipolar transistor I–V characteristics.[61]

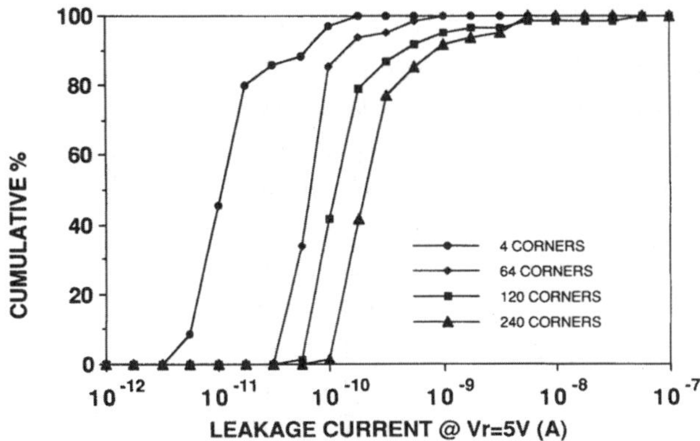

Figure 1.9 Plot of cumulative percentage of diodes versus reverse leakage for n^+–p diodes fabricated in selective epitaxial material grown at 950 °C. The diode window edges were oriented along <100>. Diode area and perimeter were 9×10^{-4} cm^2 and 0.24 cm, respectively; the number of corners was varied as shown. Median leakage current increased proportional to the number of corners.

NMOS device subthreshold leakage has been used to infer the presence of surface charge along the sidewall.[56, 58] Leakage along oxide sidewalls was 10–100× higher than a LOCOS standard, whereas leakage along nitride sidewalls was 6–8 orders of magnitude higher.[58]

Preclean Quality

The presence of the patterned mask complicates the precleaning process for selective epitaxial growth. First, formation of the window openings by plasma etch processes can damage the silicon substrate or leave etch residues on the surface; sacrificial oxidation or HCl etch may be required to remove damage or residues.[59, 60] Second, in situ precleaning of oxide masked substrates using a high-temperature bake or HCl etch may cause preferential removal of SiO_2 along the masking oxide/substrate interface.[3, 4, 62] Such "undercutting" can result in undesirable lifting of the masking layer near the window opening.

Precleaning studies have emphasized many of the defect characterization techniques mentioned in "Crystal Quality" in Section 1.3 and "Defect Density and Growth Morphology" in this section. XSEM and XTEM have been used to examine undercutting, determine reaction kinetics, and optimize the preclean process to minimize undercutting.[3, 55, 62]

Thickness

Thickness measurement techniques in selective growth can make use of the masking material surfaces as reference planes. Consequently, XSEM and surface profilometry can be used to measure the thickness of selective films. Such techniques are more practical than FTIR, SRP, or SIMS when one is dealing with the small window dimensions (up to a few tens of micrometers in size) typically found in selective growth applications.

Profilometry is attractive because of the ease of use and the minimal sample preparation involved. Profilometry has been employed to examine film growth uniformity as a function of exposed Si surface area, as a function of position across a window, and also as a function of window size.[4, 51, 54] Profilometry has been combined with FTIR thickness measurements on unpatterned wafers to determine optimum process conditions for film growth uniformity with minimal dependence on window size.[51, 63]

1.5 $Si_{1-x}Ge_x$ Epitaxial Growth

Material Considerations

Silicon and germanium form an isomorphous, single-phase solid alloy system. Bulk $Si_{1-x}Ge_x$ alloys maintain the diamond-cubic crystal structure with a lattice constant varying from 5.43 Å (Si, $x = 0$) to 5.65 Å (Ge, $x = 1$). The bulk alloy bandgap decreases from 1.11 eV (Si) to 0.67 eV (Ge) as x varies from 0 to 1. The bandgap decreases gradually (~2–4 meV/% Ge) up to $x = 0.85$, at which point the conduction band minima change from silicon-like (along $\langle 100 \rangle$ directions in reciprocal space) to germanium-like (along <111> directions in reciprocal space). The bandgap then decreases at ~14 meV/% Ge to that of Ge at $x = 1$ (Reference 64).

Epitaxial alloy film properties can differ from bulk alloy properties because the alloy film can form a pseudomorphic strained layer on the underlying Si substrate. The $Si_{1-x}Ge_x$ alloy deforms tetragonally in order to remain commensurate with the underlying Si lattice. The film remains strained, without the formation of misfit dislocations, as long as the film is thinner than a critical thickness such that the strain energy released by misfit dislocation formation is less than the energy required for dislocation formation and propagation.[65] This critical thickness decreases as the Ge content (x) increases. Metastable layers thicker than the critical thickness may grow at temperatures low enough to avoid dislocation formation. Such metastable strained layers can relax if subsequently heated to a high enough temperature.

The bandgap of the strained $Si_{1-x}Ge_x$ epitaxial film is significantly reduced from that of the bulk alloy by the strain.[66] Since control of the bandgap difference is essential for heterojunction device production, $Si_{1-x}Ge_x$ epitaxial growth must achieve reproducible, defect-free films at relatively low temperatures. Typical growth temperatures, 500–750 °C, are substantially lower than those used in typical commercial silicon epitaxial growth.

Growth temperatures also may be limited by the onset of three-dimensional growth ("islanding").[67] The onset of this growth morphology, characterized by localized epitaxial nuclei, occurs at decreasing temperatures as the Ge fraction increases.[67] The temperature at which islanding occurs also is affected by film growth methods.[67–69]

Reactor Types

Several methods have been employed for $Si_{1-x}Ge_x$ film growth. Early $Si_{1-x}Ge_x$ epitaxial growth studies utilized molecular beam epitaxy (MBE),[67] while recent efforts have emphasized various CVD techniques. One such CVD approach ("UHV/CVD") uses a UHV hot-wall chamber design to provide a low background pressure of oxidizing species.[13] Epitaxial growth can be performed at very low temperatures (400–550 °C) with this approach. This growth technique makes use of the hydrogen surface passivation provided by an ex situ HF preclean ("Preclean Quality" in Section 1.3).[33] Growth pressures are in the millitorr to torr range, and the silicon and germanium source gases are SiH_4 and GeH_4 respectively.

A second approach (rapid thermal CVD [RTCVD][68] or limited reaction processing [LRP][69]) uses a cold-wall, susceptorless reactor. The wafer is heated using lamp irradiation. The growth process can be controlled by either gas switching or thermal switching. $Si_{1-x}Ge_x$ film growth temperatures (600–900 °C) are higher than in UHV/CVD, and growth pressures are in the 1–10 torr range.[68, 69] Dichlorosilane and GeH_4 are the typical reactant gases.

A third approach uses a conventional single-wafer epitaxial reactor operating at atmospheric pressure. High-purity process gases are used, together with a

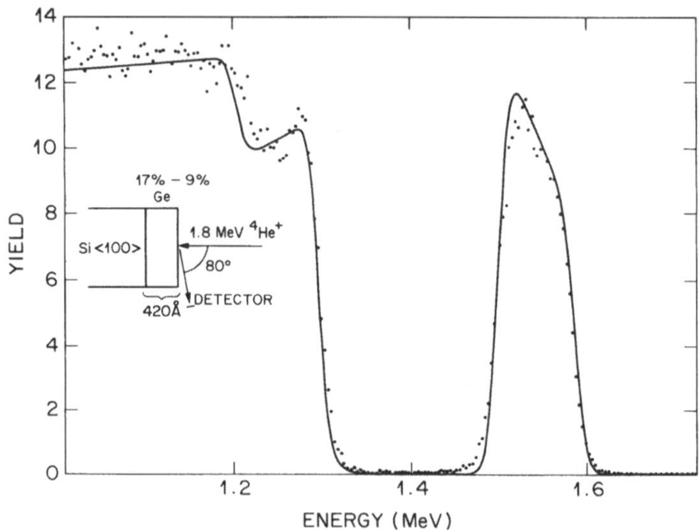

Figure 1.10   RBS spectrum of a 420 Å-thick graded $Si_{1-x}Ge_x$ film. Ge concentration varies from 17% to 9%. The points are the count data, while the solid line is a simulated spectrum used to fit the data points. The separate peak to the right is the Ge signal. (After Reference 68; reprinted by permission of M. L. Green, AT&T Bell Laboratories.)

controlled-atmosphere loadlock, to minimize the presence of contaminants. High-quality $Si_{1-x}Ge_x$ films have been grown using dichlorosilane and germane at temperatures in the 600–700 °C range.[70]

1.6 $Si_{1-x}Ge_x$ Material Characterization

Strained heteroepitaxial films of $Si_{1-x}Ge_x$ alloys require characterization beyond that of conventional silicon epitaxial layers. Additional critical parameters include Ge content, film strain, misfit dislocation density, and bandgap. The characterization challenge is increased by device requirements; for example, heterojunction bipolar transistors often require $Si_{1-x}Ge_x$ layers <500 Å thick, with extremely abrupt dopant transitions (<<50 Å/decade). Measurement of composition and thickness of such thin layers requires approaches different from those used in conventional silicon growth; assessment of such abrupt concentration changes also requires innovative techniques.

Composition and Thickness

The Ge content of $Si_{1-x}Ge_x$ films has been determined by a variety of methods, including Auger profiling,[71] SIMS,[72] X-ray diffraction,[73] and RBS,[68, 69] SIMS and Auger, both destructive techniques, require calibration against standards for best quantitation. X-ray diffraction, which is nondestructive, relies on the relation

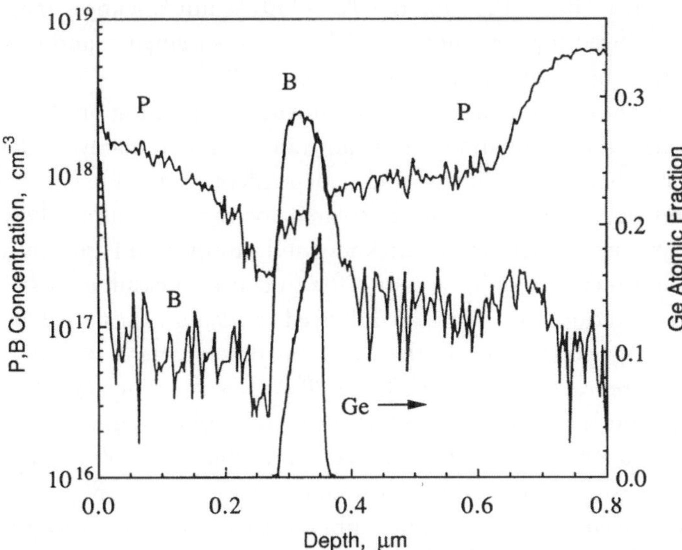

Figure 1.11 SIMS profile through a Si/Si$_{1-x}$Ge$_x$/Si HBT structure, showing the boron, phosphorus, and germanium elemental profiles. The Si$_{1-x}$Ge$_x$ layer is graded from 0 to 20%. (After Reference 72; reprinted by permission of the authors.)

between lattice parameter and Ge fraction to obtain the Ge content indirectly.[73] RBS is nondestructive and readily detects Ge in a predominantly Si matrix (Figure 1.10). RBS provides a quantitative concentration measure since the backscattering cross sections of Si and Ge are known, allowing a direct calculation of the Ge content from the backscattered spectrum. Other techniques which are sensitive to chemical or atomic properties (e.g., X-ray fluorescence) also are useful in determining Ge content of the alloy films.

Si$_{1-x}$Ge$_x$ alloy composition gradients can be introduced to grade the alloy bandgap. Such grading can be used to advantage in device applications, for example, enhancing the electron transport across the base of an NPN HBT.[9] Composition gradients can be characterized by SIMS[9, 72] (Figure 1.11) or Auger sputter profiling techniques. Typical composition gradients in a high-performance HBT are ~10%/200 Å, so that SIMS resolution is adequate. RBS also can detect composition grading[68] (Figure 1.10); with backscattering detector angle near 80°, a depth resolution of 40 Å is possible.

Film thickness also can be determined with the techniques listed for composition measurement. Thickness may be determined (destructively) by SIMS and Auger profiling from sputter-depth measurements. RBS also provides a nondestructive method for thickness measurements. As noted above, the resolution of RBS may be improved by using grazing exit-angle detection[68, 73] and by the use of simulation programs to determine best fits to the measured spectra (Figure 1.10).[68] The use of grazing exit-angle techniques is crucial for film thicknesses <1000 Å. Accuracy of

RBS has been estimated to be ~±10% for the 125–900 Å film thickness range.[74] XTEM also can be employed for (destructive) thickness measurements, and is well-suited to measuring very thin films or multilayer structures.[67, 69]

Ellipsometry recently has been demonstrated to provide information on thickness and composition of $Si_{1-x}Ge_x$ alloy films on Si substrates.[75] The refractive index of the alloy film is greater than that of silicon and is dependent on the Ge content.[76] Characteristic psi and delta curves may be calculated for the alloy films allowing ellipsometry measurements to determine thickness and composition quickly and nondestructively. Ellipsometry is quite accurate for thickness measurements (Figure 1.12). If proper calibration procedures are followed to account for native oxides and variations in the angle of incidence, the thickness measurement for 10% Ge films is estimated to be repeatable to better than ±20 Å over the range 0–800 Å.[77] Sensitivity to composition (i.e., refractive index) for 6328 Å illumination is best in the 300–550-Å thickness range (and at thicknesses of multiples of ~800 Å plus this range).[75, 76] At peak sensitivity, the germanium content of 10% Ge films can be determined to ±1%.[76] Accurate composition calibration requires comparison with one of the other methods mentioned above.

The use of ellipsometry during sputter removal of the $Si_{1-x}Ge_x$ alloy film has been used to characterize the Ge depth profile.[77] The technique was shown to be capable of resolving interfacial abruptness to within 10 Å (better than SIMS profiles of the same films). This extension of the ellipsometry technique promises a rapid method for characterizing graded-composition $Si_{1-x}Ge_x$ alloy films (e.g., for HBT fabrication).

Growth Morphology

The morphology of MBE-grown $Si_{1-x}Ge_x$ has been studied using Nomarski interference contrast in an optical microscope.[68] A change between planar growth and

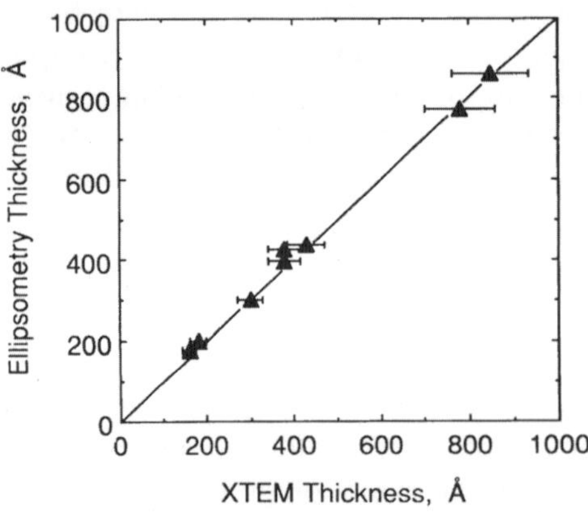

Figure 1.12　Comparison of $Si_{1-x}Ge_x$ film thickness as determined by XTEM and ellipsometry. Excellent agreement is seen between the two techniques.[76] (Reprinted by permission of the authors.)

three-dimensional "island" growth was inferred from surface roughness as the Ge fraction increased. The Ge fraction (determined using RBS) at which the transition occurred decreased significantly as the growth temperature increased; for 750 °C MBE growth, the transition occurred at ~10% Ge.[68]

Growth morphology studies in RTCVD/LRP systems have shown that island formation occurs at higher Ge fractions for a given temperature (or at higher temperatures for a given Ge content) than in MBE growth.[68, 69] For example, XTEM examination showed that 13.5% Ge films were planar when grown with SiH_2Cl_2 and GeH_4 at 900 °C.[68] As the Ge content was increased above 13.5%, the films began to exhibit undulating surfaces (unstable growth).

Lattice Strain and Critical Thickness

Lattice strain in $Si_{1-x}Ge_x$ epitaxial films has been determined using a number of techniques, including RBS[67, 68] and X-ray diffractometry (XRD).[67, 71] Both RBS and XRD can yield information about lattice spacings parallel and perpendicular to the surface and have been used to measure the tetragonal distortion of the alloy films.[67]

The measurement of critical thickness provides an interesting comparison of several different characterization methods. Measurements of critical thickness as a function of composition were performed initially using RBS, XRD, and XTEM.[67] These measurements gave much larger values for the critical thickness (particularly for small Ge fractions) than predicted by equilibrium theory. Subsequent measurements using EBIC to image misfit dislocations directly gave much smaller values of critical thickness.[71] This discrepancy was explained[71] by showing that the strain difference between commensurate and incommensurate films becomes extremely small for Ge fractions <0.3; RBS or XRD may not detect the change.

The measured critical layer thickness is thus dependent on the sensitivity of the detection method to misfit formation.[78] The estimated resolution of misfit dislocation density by XRD or RBS is expected to be ~10^4 cm^{-2}, whereas EBIC, X-ray topography, and defect etching can detect much smaller misfit densities (to ~1 cm^{-2}).[79] These higher sensitivity techniques, which provide direct evidence of misfits, are now preferred. The combination of X-ray topography and Nomarski-contrast microscopy of defect-etched surfaces has been used to determine the critical thickness for $x < 0.15$ in excellent agreement with equilibrium theory[79] (Figure 1.13). For comparison, Figure 1.13 also shows the critical thickness estimated by lower sensitivity techniques.

Methods of enhancing $Si_{1-x}Ge_x$ film stability have been studied. TEM and X-ray topography examinations demonstrate that a silicon capping layer on the $Si_{1-x}Ge_x$ film increases the film stability.[80, 81] The increased stability arises from the additional energy required to nucleate and propagate dislocations at the upper $Si/Si_{1-x}Ge_x$ interface.[81] The critical thickness thus increases in the presence of a cap layer.[68]

The use of selective growth to reduce misfit dislocation density has been studied using TEM, defect etching, and EBIC.[82, 83] A decreased misfit density in selective

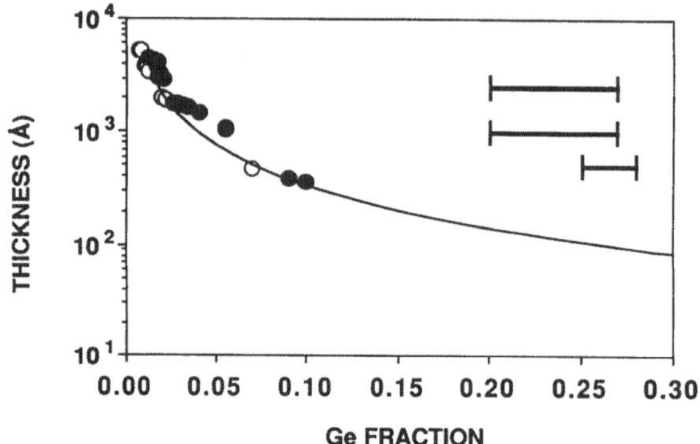

Figure 1.13 $Si_{1-x}Ge_x$ epitaxial film critical thickness versus Ge content as measured by techniques with different sensitivities. Circles are for a high-sensitivity technique (defect etching + large-area optical microscopy)[79]; open circles indicate no dislocations, while filled circles indicate relaxation. The bars show critical thickness estimates from RBS and XTEM (low-sensitivity techniques).[66] The solid line gives the theoretical boundary.[65]

films occurs because the pattern limits lateral dislocation propagation from heterogeneous nucleation points.

TEM and X-ray topography techniques have been used to examine oxygen-doped $Si_{1-x}Ge_x$ films for misfit dislocation formation.[74] Layers containing 2×10^{20} cm^{-3} oxygen were stable for thicknesses approximately twice that of the equilibrium critical thickness.

Relaxation Kinetics

The thermal stability and relaxation kinetics of $Si_{1-x}Ge_x$ films have been the subject of a large number of studies. TEM has been used extensively in these studies, and the thermally activated nature of the relaxation has made hot-stage techniques particularly appropriate. The relaxation of metastable films has been studied in situ using hot-stage TEM techniques; dislocation velocities have been determined from the observations.[84]

Bandgap Measurements

Measurement of optical absorption versus wavelength provided the original determination of the indirect bandgap in bulk alloys.[64] The bandgap in strained epitaxial films has been measured using optical absorption photocurrent.[66] These measurements confirmed that the strained-film bandgap was smaller than that of the bulk alloys.

The variation in transistor collector and base currents with temperature also has been used to extract bandgap differences between silicon and strained $Si_{1-x}Ge_x$ layers.[85] The valence-band discontinuity ΔE_v can be determined by comparing the ratio of the HBT collector current to that of a conventional homojunction device as a function of temperature[85]:

$$I_c(\text{HBT})/I_c(\text{BJT}) \approx \exp(\Delta E_v/kT)$$

The *total* energy gap difference ΔE_g can be found from the temperature variation of the ratio of the HBT base current to the homojunction collector current. Modulation doping measurements as a function of temperature[11] suggest that the conduction band discontinuity ΔE_c is small, so that $\Delta E_g \approx \Delta E_v$. The measured bandgap-difference dependence on Ge content in unrelaxed films agrees well with optical absorption measurements and with values predicted by band calculations.[85]

Interfacial Abruptness and Outdiffusion

Lattice-imaging XTEM can measure the interfacial abruptness of the heterostructure on an atomic scale. However, XTEM is most sensitive to the change in the Ge content and does not indicate the change in dopant concentration across a heterojunction. In addition, XTEM samples a small area ($\sim$0.1 $\times$ 1–100 μm), SIMS and other sputtering techniques sample larger areas ($\sim$300 $\times$ 300 μm), but depth resolution is limited (to $\sim$50 Å/decade for boron[43]).

Electrical techniques have been used for improved resolution of heterojunction abruptness. Outdiffusion of boron from the $Si_{1-x}Ge_x$ base of an NPN HBT will alter the band offset of the heterojunction. Modeling of this phenomenon has been used together with careful measurements of the band offset (see previous section, "Bandgap Measurements") to estimate the outdiffusion of boron from RTCVD-deposited $Si_{1-x}Ge_x$ films.[86] This technique is estimated to be able to detect boron diffusion profiles with characteristic diffusion lengths as small as 20 Å.[86]

Two-dimensional hole gases may be formed by means of modulation doping in Si/$Si_{1-x}Ge_x$ heterostructures. Measurements of both hole-gas mobility and lowtemperature magnetoresistive (Shubnikov-de Haas) effects have been combined to infer the heterostructure abruptness.[86, 87] Overall interface abruptness in RTCVD-deposited $Si_{1-x}Ge_x$ films has been estimated at $\leq$10 Å using this technique. A boron dopant concentration gradient of nine orders of magnitude in less than 100 Å has been estimated using this method on UHV/CVD-deposited films.[87]

RBS has been used to study the Ge outdiffusion from $Si_{1-x}Ge_x$ layers during postdeposition processing.[68] The Ge diffusion coefficient in silicon has been determined as a function of temperature, and a 33-Å diffusion length for a 950 °C, 1-h anneal has been estimated.

Impurity Profiles

Because of its relatively high sensitivity and its ability to examine layers <0.1 μm thick, SIMS has been the impurity profiling technique of choice for $Si_{1-x}Ge_x$/Si heterostructures. SIMS has been applied to HBT dopant profiling[72, 85] (Figure 1.11) and also to the characterization of oxygen in the alloy films.[68, 74]

1.7 Summary

A broad array of characterization techniques have been applied to silicon-based epitaxial films. This chapter has presented a number of process and material characterization problems and has given the most widely used approaches for each problem. The examples have attempted to illustrate the typical situations in which a given technique is applicable. It should be noted that many characterization tasks (film thickness, carrier concentration, Ge content in $Si_{1-x}Ge_x$ films, etc.) have multiple approaches, allowing the user considerable flexibility.

A close relationship exists between characterization capability and process improvement. The advances in silicon epitaxial growth over the past decade would not have been achieved without access to sophisticated analytical equipment. An illustration of this point is the development of low-temperature chemical vapor deposition techniques for silicon and $Si_{1-x}Ge_x$ epitaxy, resulting from a tremendous increase in our knowledge of surface cleaning. The advance in cleaning methods has been brought about by detailed analyses of the silicon surface during and after treatment. Such study has led to the reproducible chemical vapor deposition of epitaxial silicon at temperatures unheard of ten years ago.

These sophisticated low-temperature growth processes, supported by the array of advanced analytical equipment now available, have opened up the remarkable possibility of mass production of silicon-based heterostructures. Continued advances in this direction raise the hope that silicon, too, will be the material of the future.

Acknowledgments

The author would like to thank Drs. J. E. Turner (Hewlett-Packard), M. L. Green (AT&T Bell Laboratories), B. S. Meyerson (IBM T. J. Watson Research Center), T. O. Sedgwick (IBM T. J. Watson Research Center), and J. C. Sturm (Princeton University) for generously giving permission to reprint figures as illustrations for this chapter. He also thanks Dr. M. Racanelli and L. K. Garling of Motorola, Inc., for their contributions to this chapter and acknowledges Drs. S. Thomas and P. Gill of Motorola, Inc., for their continued support and encouragement.

References

1   R. R. Troutman. *Latchup in CMOS Technology: The Problem and Its Cure.* Kluwer Academic Publishers, Boston, 1986.

2   *BiCMOS Technology and Applications.* (A. R. Alvarez, Ed.) Kluwer Academic Publishers, Boston, 1989.

3   J. O. Borland and C. I. Drowley. *Solid State Technology.* **28** (8), 141, 1985.

4   A. Ishitani, H. Kitajima, K. Tanno, H. Tsuya, N. Endo, N. Kasai, and Y. Kurogi. *Microelectronic Eng.* **4**, 3, 1986.

5   S. S. Wong, D. R. Bradbury, D. C. Chen, and K.-Y. Chiu. *1984 IEDM Tech. Digest*. p. 634.

6   N. C. C. Lu, T. V. Rajeevakumar, G. B. Bronner, B. Ginsberg, B. J. Machesny, and E. J. Sprogis. *1988 IEDM Tech. Digest*. p. 588.

7   T. Yamada, S. Samata, H. Takato, Y. Matsushita, K. Hieda, A. Nitayama, F. Horiguchi, and F. Masuoka. *1989 IEDM Tech. Digest*. p. 35.

8   R. P. Zingg, B. Hofflinger, and G. W. Neudeck. *IEEE Electron Dev. Lett.* **11**, 9, 1990.

9   G. L. Patton, J. H. Comfort, B. S. Meyerson, E. F. Crabbe, G. J. Scilla, E. de Fresart, J. M. C. Stork, J. Y.-C. Sun, D. L. Harame, and J. N. Burghartz. *IEEE Electron Dev. Lett.* **11**, 171, 1990.

10 K. Imai, T. Yamazaki, T. Tashiro, T. Tatsumi, T. Niino, N. Aizaki, and M. Nakamae. *Proceedings*. 1990 Bipolar Circuit Technical Meeting, p. 90.

11 R. People, J. C. Bean, D. V. Lang, A. M. Sergent, H. L. Stormer, K. W. Wecht, R. T. Lynch, and K. Baldwin. *Appl. Phys. Lett.* **45**, 1231, 1984.

12 T. P. Pearsall, H. Temkin, J. C. Bean, and S. Luryi. *IEEE Electron Device Lett.* **7**, 330, 1986.

13 B. S. Meyerson. *Appl. Phys. Lett.* **48**, 797, 1986.

14 W. R. Burger and R. Reif. *J. Appl. Phys.* **62**, 4255, 1987.

15 F. W. Smith and G. Ghidini. *J. Electrochem. Soc.* **129**, 1300, 1982.

16 G. Ghidini and F. W. Smith. *J. Electrochem. Soc.* **131**, 2924, 1984.

17 P. D. Agnello and T. O. Sedgwick. *Proceedings*. 11th International Conf. on Chem. Vapor Deposition. Electrochemical Society, Pennington, NJ, 1990, p. 247.

18 T. O. Sedgwick and P. D. Agnello. *Proceedings*. 11th International Conf. on Chem. Vapor Deposition. Electrochemical Society, Pennington, NJ, 1990, p. 254.

19 "Test Method for Stacking Fault Density of Epitaxial Layers of Silicon by Interference-Contrast Microscopy". *Annual Book of ASTM Standards*. F 522–88. American Society for Testing and Materials, Philadelphia, 1988.

20 T. Brown. *Semiconductor International*. April 1990, p. 134.

21 D. J. Ruprecht, L. G. Hellwig, and J. A. Rossi. In *Semiconductor Fabrication: Technology and Metrology, ASTM STP 990*. (D. C. Gupta, Ed.) American Society for Testing and Materials, Philadelphia, 1989, p. 87.

22 F. Secco d'Aragona. *J. Electrochem. Soc.* **119**, 948, 1972

23 M. W. Jenkins. *J. Electrochem. Soc.* **124**, 758, 1977.

24 D. G. Schimmel. *J. Electrochem. Soc.* **126**, 479, 1979.

25 "Test Method for Crystallographic Perfection of Epitaxial Deposits of Silicon by Etching Techniques." *Annual Book of ASTM Standards.* F 80–88a. American Society for Testing and Materials, Philadelphia, 1988.

26 K. V. Ravi. *Imperfections and Impurities in Semiconductor Silicon.* J. Wiley and Sons, New York, 1981, Chapts. 4 and 5.

27 B. O. Kolbesen and H. P. Strunk. In *VLSI Electronics: Microstructure Science.* Vol. 12. (N. G. Einspruch and H. Huff, Eds.) Academic Press, New York, 1985, Chapt. 4.

28 M. P. Scott, L. Caubin, D. C. Chen, E. R. Weber, J. Rose, and T. Tucker. *Mat. Res. Soc. Symp. Proc.* **36**, 37, 1985.

29 F. Hottier and R. Cadoret. *J. Crystal Growth.* **61**, 245, 1983.

30 Y. Kobayashi, Y. Shinoda, and K. Sugii. *Jpn. J. Appl. Phys.* **29**, 1004, 1990.

31 V. A. Burrows, Y. J. Chabal, G. S. Higashi, K. Raghavachari, and S. B. Christman. *Appl. Phys. Lett.* **53**, 998, 1988.

32 N. Hirashita, M. Kinoshita, I. Aikawa, and T. Ajioka. *Appl. Phys. Lett.* **56**, 451, 1990.

33 B. S. Meyerson, F. J. Himpsel, and K. J. Uram. *Appl. Phys. Lett.* **57**, 1034, 1990.

34 B. Anthony, T. Hsu, L. Breaux, R. Qian, S. Banerjee, and A. Tasch. *J. Electronic Materials.* **19**, 1027, 1990.

35 T. Takahagi, A. Ishitani, H. Kuroda, Y. Nagasawa, H. Ito, and S. Wakao. *J. Appl. Phys.* **68**, 2187, 1990.

36 "Test Method for Thickness of Epitaxial Layers of Silicon on Substrates of the Same Type by Infrared Reflectance". *Annual Book of ASTM Standards.* F 95–89. American Society for Testing and Materials, Philadelphia, 1989.

37 P. F. Cox and A. F. Stalder. *J. Electrochem. Soc.* **120**, 287, 1973.

38 "Test Method for Measuring Resistivity Profiles Perpendicular to the Surface of a Silicon Wafer Using a Spreading Resistance Probe." *Annual Book of ASTM Standards.* F 672–88. American Society for Testing and Materials, Philadelphia, 1988.

39 S. M. Hu. *J. Appl. Phys.* **53**, 1499, 1982.

40 H. L. Berkowitz, D. M. Burnell, R. J. Hillard, R. G. Mazur, and P. Rai-Choudhury. *Solid-State Electronics.* **33**, 773, 1990.

41 "Test Method for Thickness of Epitaxial or Diffused Layers in Silicon by the Angle Lapping and Staining Technique." *Annual Book of ASTM Standards.* F 10–88. American Society for Testing and Materials, Philadelphia, 1988.

42 B. McDonald and A. Goetzberger. *J. Electrochem. Soc.* **109**, 141, 1962.

43 J. E. Turner, J. Amano, C. M. Gronet, and J. F. Gibbons. *Appl. Phys. Lett.* **50**, 1601, 1987.

44 C. M. Gronet, J. C. Sturm, K. E. Williams, J. F. Gibbons, and S. D. Wilson. *Appl. Phys. Lett.* **48**, 1012, 1986.

45 "Method for Measuring Resistivity of Silicon Slices with a Colinear Four-Probe Array". *Annual Book of ASTM Standards.* F 84–8. American Society for Testing and Materials, Philadelphia, 1988.

46 M. Wong, R. Reif, and G. R. Srinivasan. *IEEE Trans. Electron Devices.* **32**, 89, 1985.

47 P. J. Severin and G. J. Poodt. *J. Electrochem. Soc.* **119**, 1384, 1972.

48 W. C. Johnson and P. T. Panousis. *IEEE Trans. Electron Devices.* **18**, 965, 1971.

49 "Test Methods for Measuring Resistivity and Hall Coefficient and Determining Hall Mobility in Single Crystal Semiconductor". *Annual Book of ASTM Standards.* F 76–88. American Society for Testing and Materials, Philadelphia, 1988.

50 R. Baron, G. Shifrin, O. J. Marsh, and J. W. Mayer. *J. Appl. Phys.* **40**, 3702, 1969.

51 C. I. Drowley. *Proceedings.* 10th Internationall Conf. on Chem. Vapor Deposition. Electrochemical Society, Pennington, NJ, 1987, p. 418.

52 C. Galewski and W. G. Oldham. *Tech. Digest 1988 Symp. VLSI Technology,* p. 79.

53 L. Jastrzebki, J. F. Corboy, J. T. McGinn, and R. Pagliaro. *J. Electrochem. Soc.* **130**, 1571, 1983.

54 R. Pagliaro, Jr., J. F. Corboy, L. Jastrzebski, and R. Soydan. *J. Electrochem. Soc.* **134**, 1235, 1987.

55 T. Y. Hsieh, K. H. Jung, D. L. Kwong, and S. K. Lee. *J. Electrochem. Soc.* **138**, 1188, 1991.

56 A. Stivers, C. Ting, and J. Borland. *Proceedings.* 10th International Conf. on Chem. Vapor Deposition. Electrochemical Society, Pennington, NJ, 1987, p. 389.

57 C. I. Drowley, G. A. Reid, and R. Hull. *Appl. Phys. Lett.* **52**, 546, 1988.

58 N. Endo, N. Kasai, A. Ishitani, H. Kitajima, and Y. Kurogi. *IEEE Trans. Electron Devices.* **33**, 1659, 1986.

59 D. Harame, B. Ginsberg, M. Arienzo, S. Mader, and M. D'Agostino. *Solid-State Electronics.* **30**, 907, 1987.

60 W. A. Klaasen and G. W. Neudeck. *IEEE Trans. Electron Devices.* **37**, 273, 1990.

61 J. W. Siekkinen, W. A. Klaasen, and G. W. Neudeck. *IEEE Trans. Electron Devices.* **35**, 1640, 1988.

62 S. T. Liu, L. Chan, and J. Borland. *Proceedings.* 10th Int'l Conf. on Chem. Vapor Deposition. Electrochemical Society, Pennington, NJ, 1987, p. 428.

63 C. I. Drowley and M. L. Hammond. *Solid State Technology.* May 1990, p. 135.

64 R. Braunstein, A. R. Moore, and F. Herman. *Phys. Rev.* **109**, 695, 1958.

65 J. W. Matthews and A. E. Blakeslee. *J. Cryst. Growth.* **27**, 118, 1974.

66 D. V. Lang, R. People, J. C. Bean, and A. M. Sergent. *Appl. Phys. Lett.* **47**, 1333, 1985.

67 J. C. Bean, L. C. Feldman, A. T. Fiory, S. Nakahara, and I. K. Robinson. *J. Vac. Sci. Technol.* **A2**, 436, 1984.

68 M. L. Green, B. E. Weir, D. Brasen, Y. F. Hsieh, G. Higashi, A. Feygenson, L. C. Feldman, and R. L. Headrick. *J. Appl. Phys.* **69**, 745, 1991.

69 J. L. Hoyt, C. A. King, D. B. Noble, C. M. Gronet, J. F. Gibbons, M. P. Scott, S. S. Laderman, S. J. Rosner, K. Nauka, J. Turner, and T. I. Kamins. *Thin Solid Films.* **184**, 93, 1990.

70 T. I. Kamins and D. J. Meyer. *Ext. Abs. Electrochem. Soc.,* Vol. 90–2. Electrochemical Society, Pennington, NJ, 1990, p. 685.

71 Y. Kohama, Y. Fukuda, and M. Seki. *Appl. Phys. Lett.* **52**, 380, 1988.

72 J. C. Sturm, E. J. Prinz, and C. W. Magee. *IEEE Electron Device Lett.* **12**, 303, 1991.

73 C. A. King, J. L. Hoyt, C. M. Gronet, J. F. Gibbons, M. P. Scott, and J. E. Turner. *IEEE Electron Dev. Lett.* **10**, 52, 1989.

74 D. B. Noble, J. L. Hoyt, W. D. Nix, J. F. Gibbons, S. S. Laderman, J. E. Turner, and M. P. Scott. *Appl. Phys. Lett.* **58**, 1536, 1991.

75 T. I. Kamins. *Electronics Lett.* **27**, 451, 1991.

76 M. Racanelli, C. I. Drowley, N. D. Theodore, R. B. Gregory, H. G. Tompkins, and D. J. Meyer. *Appl. Phys. Lett.* **60**, 2225, 1992.

77 G. M. W. Kroesen, G. S. Oehrlein, E. de Fresart, and G. J. Scilla. *Appl. Phys. Lett.* **60**, 1351, 1992.

78 I. J. Fritz. *Appl. Phys. Lett.* **51**, 1080, 1987.

79 D. C. Houghton, C. J. Gibbings, C. G. Tuppen, M. H. Lyons, and M. A. G. Halliwell. *Appl. Phys. Lett.* **56**, 460, 1990.

80 R. Hull and J. C. Bean. *Appl. Phys. Lett.* **55**, 1900, 1989.

81 D. B. Noble, J. L. Hoyt, J. F. Gibbons, M. P. Scott, S. S. Laderman, S. J. Rosner, and T. I. Kamins. *Appl. Phys. Lett.* **55**, 1978, 1989.

82 D. B. Noble, J. L. Hoyt, C. A. King, J. F. Gibbons, T. I. Kamins, and M. P. Scott. *Appl. Phys. Lett.* **56**, 51, 1990.

83 E. A. Fitzgerald, Y.-H. Xie, D. Brasen, M. L. Green, J. Michel, P. E. Freeland, and B. E. Weir. *J. Elect. Mat.* **19**, 949, 1990.

84 R. Hull, J. C. Bean, D. J. Werder, and R. E. Leibenguth. *Appl. Phys. Lett.* **52**, 1605, 1988.

85 C. A. King, J. L. Hoyt, and J. F. Gibbons. *IEEE Trans. Electron Dev.* **36**, 2093, 1989.

86 J. C. Sturm, P. M. Garone, E. J. Prinz, P. V. Schwartz, and V. Venkataraman. *Proceedings.* 11th Internationall Conf. on Chem. Vapor Deposition. Electrochemical Society, Pennington, NJ, 1990, p. 295.

87 B. S. Meyerson, P. J. Wang, F. K. LeGoues, G. L. Patton, J. H. Comfort, E. F. Crabbe, and G. J. Scilla. *Proceedings.* 11th International Conf. on Chem. Vapor Deposition. Electrochemical Society, Pennington, NJ, 1990, p. 229.

Polysilicon Conductors

YALE STRAUSSER

Contents

2.1 Introduction

Single-crystal silicon, silicon dioxide, and polycrystalline silicon (polysilicon) are the three materials that enabled the semiconductor revolution. The existence of these three forms of the same very abundant material, with the properties to provide a semiconductor, insulator, and conductor, respectively, is the cornerstone of the technology. They are also the most widely used materials in semiconductor integrated circuit (IC) manufacturing. This is probably no surprise to most people in the case of single-crystal silicon and silicon dioxide. However, the many uses of polysilicon are not as widely known. Silicon dioxide's importance stems partly from the ease with which it can be grown by thermal oxidation on a silicon surface and its stability during subsequent thermal treatment. Polysilicon's importance stems partly from the ease with which it can be deposited on silicon or silicon-based dielectrics and from its compatibility with subsequent high-temperature processing steps.

Polysilicon is widely used in the manufacture of ICs as the material for MOS transistor gates, as an interconnect conductor material, as both emitter and base in bipolar integrated circuits, as capacitor electrodes, as resistors, as a dopant source, as an impurity getter when deposited on the wafer backside, and as charge-storage layers in various kinds of memories. It is also widely used as a semiconductor in solar cells and as the active layer in thin-film transistors in various display devices.

In each of these applications, certain materials properties of the polysilicon are critical to the operation of the device.

A variety of characterization tools have been used at each of the many steps of the manufacturing process to monitor the properties of the polysilicon and to solve processing problems. Different properties must be monitored at different steps in the process to ensure the correct values of the critical properties in the final product. This chapter discusses a recommended set of characterization tools and their use in measuring materials properties which are of concern in a generic process. The chapter is organized as a typical process flow: deposition, doping, patterning, and effects of subsequent processing. Although in a particular application these steps may be intermixed (e.g., in situ doping during deposition) or omitted, this is a good average process history. A discussion of these processing steps, the material properties that are important, and some of the typical problems that arise are used to illustrate the application of characterization tools to the solution of process problems.

The material properties of polysilicon which are important in most of its applications are relatively few. Polysilicon is generally used for its electrical transport properties, although there are applications in which its optical properties are important (such as in thin-film transistors for displays and solar cells). Resistivity is the most frequently measured parameter in process control. This is key in a polysilicon gate in a field effect transistor, a polysilicon or polycide interconnect line in any IC, in a solar cell, or in a thin-film transistor in a display device. In order to control resistivity in the final product, one must control many other parameters throughout the process flow. These include parameters as conceptually simple to measure as film thickness, or as complicated as dopant incorporation and activation in the silicon lattice.

Entire books (e.g., References 1 and 2) have been devoted to discussions of the processing, properties, and applications of polysilicon. This chapter addresses these issues in a more limited manner as a background to the discussion of the appropriate use of characterization techniques in solving materials problems.

2.2 Deposition

The deposition of polysilicon is usually done by chemical vapor deposition (CVD) using silane, disilane, or dichlorosilane as the silicon source gas. The source gas is thermally decomposed on the substrate surface, or sometimes partially in the boundary layer of the flowing gas, leaving silicon atoms on the surface and gas-phase reaction products to be carried away in the gas flow. The source gas generally is a mixture with a carrier gas, such as argon. The deposition may be done at atmospheric pressure (APCVD) or at lower pressures (LPCVD). Deposition is typically done in a long quartz tube capable of holding many wafers at a time, although there is increasing interest in single-wafer deposition using a rapid thermal (RTCVD) process. Depending on pressure, flow rate, and—most importantly—substrate temperature, the silicon may be deposited as amorphous silicon or as polysilicon.

When silicon is deposited in amorphous form, subsequent thermal processing usually leads to crystallization.[3]

Surface Preparation

During deposition, the silicon atoms on the surface diffuse across the surface until a nucleation site is found. The nucleation site may be a random encounter with another silicon atom forming a silicon dimer and beginning a nucleus for crystal growth, or it may be a chemical, structural, or electronic impurity site—adsorbed foreign atoms on a surface, ledges, or steps on a crystalline surface, trapped charge sites on insulator surfaces, etc. The nucleation site density is partly controlled by controlling the densities of impurity sites. The as-deposited grain size of polysilicon films is a function of the nucleation site density. Thus, the first point in the process flow where surface or microstructural characterization tools are important is in the control of the substrate surface conditions as deposition is ready to begin.

Foreign atom concentrations on surfaces prior to deposition are best measured by Auger electron spectroscopy (AES), X-ray photoelectron spectroscopy (XPS), static secondary ion mass spectroscopy (static SIMS), or total internal reflection X-ray fluorescence (TXRF). These techniques are described in the lead volume of this series, the *Encyclopedia of Materials Characterization*. For most applications in this area a detection sensitivity of 1 part in 10^4 is sufficient, allowing the use of AES or XPS. When a higher sensitivity is needed, static SIMS or TXRF may be used, depending on the foreign-element-substrate combination.

There are well-established, well-documented cleaning procedures for silicon which, when used under proper control, result in atomically clean silicon substrate surfaces. Likewise, there are well-known procedures for cleaning SiO_2.[4, 5] These are the two materials onto which polysilicon is typically deposited. There are, however, many ways in which the substrate cleaning procedures can go out of control. Inadequate cleaning typically results in haze on the wafer surface following deposition—diffuse rather than specular reflectance from the as-deposited surface. Upon the appearance of haze in a normally smoothly running process, a careful analysis after each step of the cleaning process will generally turn up an impurity residue. Three techniques are presently successfully applied in renewed efforts to control semiconductor wafer cleaning[6, 7]: static SIMS, TXRF, and a wet-chemistry-based technique (which involves vapor phase hydrofluoric acid dissolution of the surface oxide layer from the wafer, collection of the droplets so produced, followed either by flameless atomic absorption of the solution or by drying of the solution and analysis of the residue it leaves by TXRF). Depending on the impurity and the surface, one of these techniques will generally provide sensitivity limits of 10^{10}–10^{12} atoms/cm^2. This is equivalent to 10^{15}–10^{18} atoms/cm^3 in a homogeneously distributed bulk impurity, or 20 ppm to 20 ppb in silicon, or 2 parts in 10^5 to 2 parts in 10^8.

Ledges and steps on crystalline surfaces are typically characterized for morphology by transmission electron microscopy (TEM) or scanning tunneling microscopy

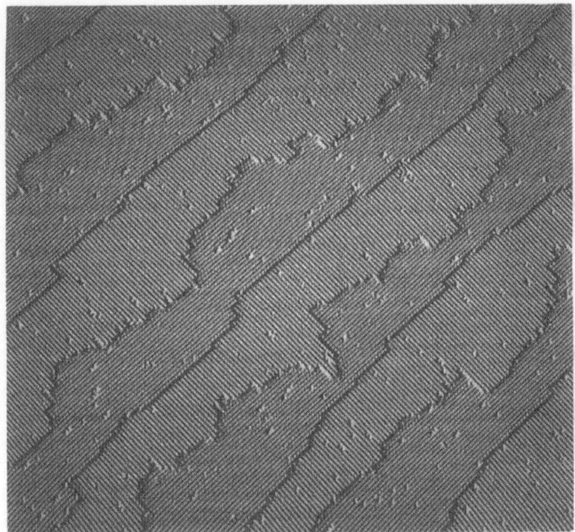

Figure 2.1   Image of a Si crystal surface magnified approximately 1 million times, taken with an STM. The surface is composed of single-atomic-layer steps that step down from the upper left to the lower right of the figure. Rows of atoms consisting of Si dimers are clearly visible on each terrace. Because of the Si crystal structure the rows are perpendicular to each other on adjacent terraces. The differing roughness of alternating terrace edges is caused by the differing dimer row termination possible in the two different dimer orientations.

(STM). The use of replica techniques and plan-view TEM has long been the preferred method for determining surface ledge and step densities when large steps are present. When single-atom steps are important, it becomes necessary to coat the surface and use cross-sectional TEM. Even with today's atomic resolution TEMs and precise sample preparation techniques, this is a difficult measurement to make. The ease of use of the recently developed STM, which can resolve single-atom height differences on doped silicon, has made it the tool of choice for characterizing atomic ledge and step densities and even mono-atomic vacancies on surfaces.

Nucleation and Growth

Figures 2.1 and 2.2 show STM images of silicon films during the early stages of nucleation and growth on a single silicon crystal surface by molecular beam epitaxial (MBE) growth.[8] Although this growth technique is not typically used to produce polycrystalline silicon, this example is representative of the characterization of the nucleation and growth process that can be accomplished through the use of STM. Figure 2.1 is an STM image of a clean vicinal silicon (001) surface before any silicon deposition. Because this sample surface was cut at approximately 0.1° to the true (001) surface, the (001) surface plane steps are compressed closely together with only narrow exposed regions on each plane. Many vacant surface atom positions are visible on each plane. These vacant positions and the ledges

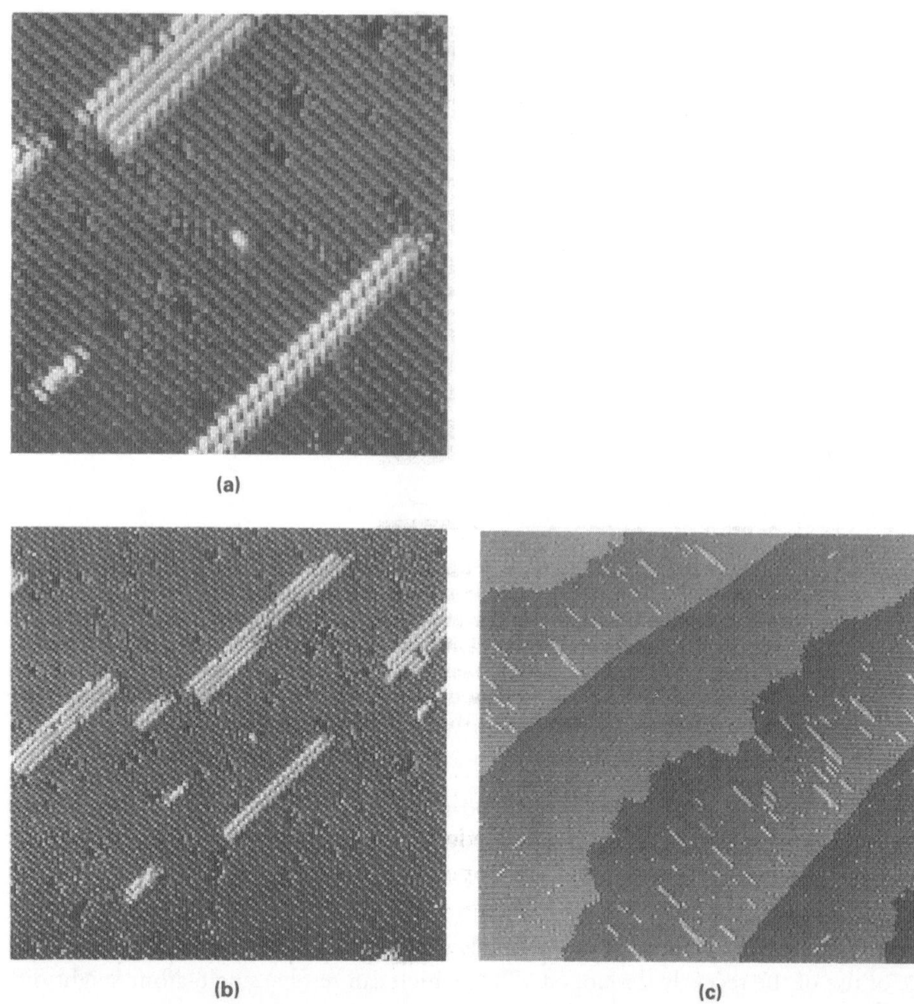

Figure 2.2 STM scan of an area on Si(001) after 0.1 monolayer of Si had been deposited. The white patches are anisotropic islands formed by Si adatoms. The smallest consists of two Si atoms. Vacancies in the substrate are visible as dark "holes." Imaged areas are (*a*) 240 × 240 Å, (*b*) 600 × 600 Å, and (*c*) 5000 × 5000 Å. The terraces are alternatingly populated by and denuded of islands because surface diffusion is anisotropic.

bounding the exposed planes are the first sites where silicon atoms will be retained. Figure 2.2*a* is a high-magnification image of a 240 × 240 Å area of one such surface after 0.1 monolayer of silicon has been deposited on it at 320 °C. Newly deposited epitaxial silicon starting to grow in the next surface layer can now be seen (bright areas) as well as some remaining vacant atom positions (dark area) Figure 2.2*b* is a lower magnification image (covering 600 × 600 Å) of the same surface as Figure 2.2*a*. The anisotropic growth shapes of the silicon islands become very clear. This has been hypothesized to be due to anisotropy in the accommodation of new silicon atoms on the sides of growing islands and to thermodynamic factors resulting from the energy required to form an edge in each of the two epitaxial directions. Figure 2.2*c* shows an even lower magnification image of a 5000 × 5000 Å region of the same sample and the preferential growth of new islands on planes with a particular orientation. The STM is only now being extensively applied to this problem, but it is clear that it has much to offer to the understanding of the nucleation and growth process and in identifying nucleation and growth anomalies.

Firmer conclusions can be reached in solving nucleation and growth problems with the help of STM images, but the STM requires a conducting surface to produce images. If the substrate surface is a dielectric film, such as SiO_2, the atomic force microscope (AFM) must be used instead. The AFM doesn't quite match the STM in spatial resolution yet, though features as small as single atoms have been resolved.

The atomic scale processes involved in the nucleation and growth of silicon films are presently receiving renewed attention, and STM provides previously unimaginable detail. A second much improved technique is spectroscopic ellipsometry (SE). New hardware configurations and, more importantly, new data-processing capabilities have given SE the power to provide in situ measurements of nucleation and growth with the sensitivity necessary to determine composition, coverage fraction, nucleus shape, and other vital information from the very early, submonolayer stage of the process through bulk film growth to the completion of deposition.

A good example of the application of SE to silicon nucleation and initial growth is in a study of magnetron-sputtered silicon growing on a thermal oxide film on c–silicon.[9] Again, this is not a typical polysilicon growth situation, but it illustrates well the breadth of information available from SE. Real time, in situ spectroscopic ellipsometry data were acquired over the energy range from 1.5 to 4.5 eV at 14.5 s intervals with real-time data processing and display during film deposition.

In the initial nucleation phase, it was shown that 15% of the surface was covered with nuclei with a mean height of 15 Å. The nucleation density was about 5×10^{12}/cm^2. At this point, a full monolayer coverage then grew to completeness over the remaining oxide surface, whereas the mean nucleus height grew to 24 Å. Once the monolayer coverage was complete, bulk film growth took place, retaining the 24 Å roughness of the original nuclei. This, plus similar data from other measurements, suggests that the columnar growth typically seen in CVD growth of polysilicon starts from similar tall nuclei but continues to grow without the monolayer filling phase. Acquiring this type of information during polysilicon process development would then permit monitoring of the SE output in real time as a process control monitor with very high sensitivity to growth variables. This has been done, for example, in MBE growth of AlGaAs on GaAs.[11]

As does STM, SE has the capability of providing significant advances in our fundamental understanding of the nucleation and growth processes. Equally valuable, and perhaps more so in the present context, is the role it will play in process problem-solving and eventually in in situ process monitoring.

Grain size analysis After a film has been deposited, the best method for determining the nucleation site density and resulting grain size is cross-sectioning it for microscopic examination. Selection of cross-sectioning techniques depends upon the substrate material and structure and on the type of microscopy to be used. The simple structure of a polysilicon film grown directly on a single-crystal silicon wafer might best be cross-sectioned by scribing the wafer on its back, cleaving the wafer, and directly observing the cross section. Optical microscopy will generally be inadequate for examining as-deposited polysilicon films because the grain size is typically in the range 50–1000 Å. Scanning electron microscopy (SEM) has the resolution required to see these small grains, but will only provide grain–grain differentiation if the grain boundaries are delineated with a chemical etchant. Once the sample is prepared, it may have to be coated before examination to avoid charging while being probed by the electron beam in the SEM, depending on its resistivity.

More detail is available through TEM cross-sectional analysis. The TEM can give a higher spatial resolution and provide information on defects within the grains and on the orientation of the grains. Sample preparation requires more time and effort. For TEM analysis, a thin section must be made from the sample, thin enough that the electrons can pass through the sample without suffering multiple scattering collisions with the sample material. For silicon this means that the sample, in the area being analyzed, must be less than 0.5 μm thick for analysis by a 100-keV electron beam. Increasing the analysis beam energy to 200 keV allows the analyzed area to have a thickness of up to approximately 1.1 μm.

A series of TEM images is shown in Figure 2.3, including cross-sectional images (2.3*a*) of an as-deposited film and a similar film after doping. Cross-sectional views such as these show that grain growth occurs as the growth proceeds, and they give a limited sampling of the variation of surface morphology with grain growth. They also give an indication of grain size, but since only one slice from each grain is seen, they don't give any quantitative information on grain diameters or statistically valid grain size distributions. Since these grains are columnar, a plan-view TEM image, such as the one in Figure 2.3*b*, of the as-deposited polysilicon, or in Figure 2.3*c*, of the doped polysilicon is a more useful tool for the accurate measurement of the grain size distribution because the entire grain outline can be seen and more grains can be included in one image, improving the statistics. Depending on which thin slice of the original film is left when the thin-foil TEM sample preparation is finished, the small grains frequently present at the bottom of the film may not be present in the plan-view image. These small grains are generally not a factor in any problem helped by grain size analysis. The as-deposited polysilicon film is so highly defected with twins and other crystalline defects that it is nearly impossible to define the grain boundaries for grain size analysis. Here it is only possible to say that the grains are approximately 500 Å in diameter. However, in the case of a doped sample, although there are still a few twins left in the grains, one can straight-forwardly identify the grain boundaries and apply grain size analysis software to

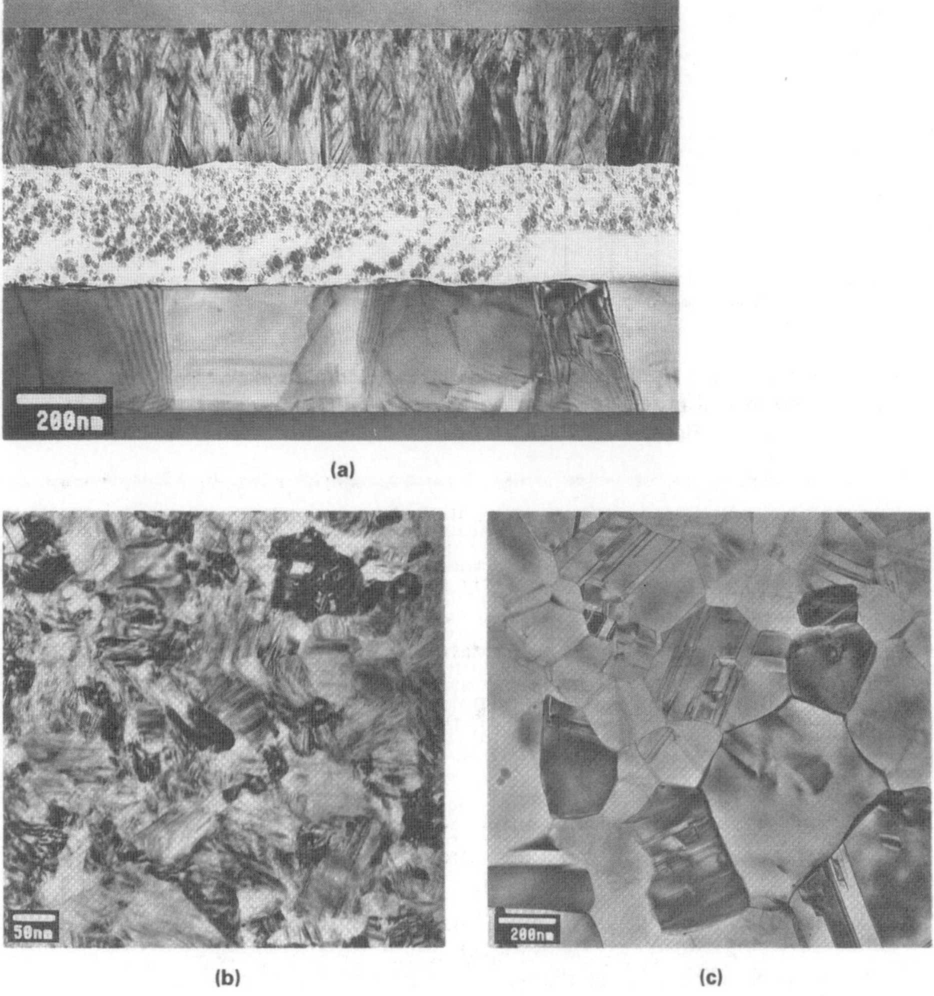

Figure 2.3   (*a*) A cross-sectional TEM image of a sample made by gluing, face-to-face, an as-deposited polysilicon film, on single-crystal silicon, to a film taken from the same deposition but which has been diffusion-doped with P. The doped film is on the bottom. The bright band through the middle of the figure is the glue. (*b*) A plan-view TEM image of the as-deposited film whose cross section is seen in the top of (*a*). (*c*) A plan-view TEM image of the doped film whose cross section is seen in the bottom of (*a*).

such an image. The distribution of grain sizes determined from this image with the help of one of the many grain-size-analyzing computer programs available is shown in Figure 2.4 as a histogram of the size distribution of the 46 grains in this photograph. For a good, statistically significant grain size distribution, one should include at least 100 grains to determine a distribution. In this case, data from additional TEM images would be required.

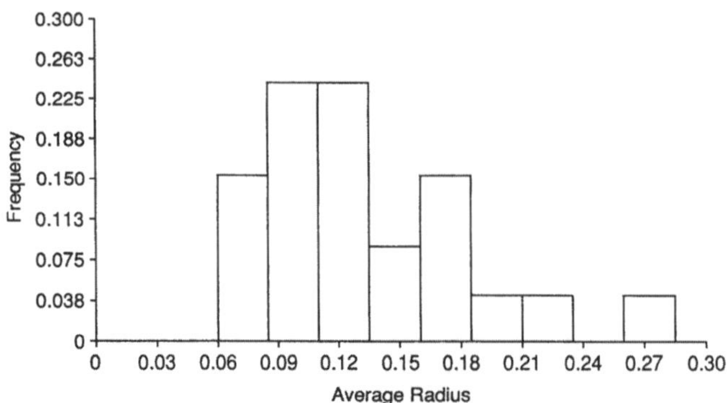

Figure 2.4   Grain size histogram of the grain size distribution of the film shown in Figure 2.3c.

Grain size analysis is not extensively done on as-deposited films. At this stage of processing, the grains are usually very small and highly defected with dislocations, twins, and other crystalline defects. After doping and other annealing steps, when the grains have grown to near their final size, TEM is more extensively used in relating grain morphology to resistivity variations, problems in doping kinetics, problems in etching, or problems in oxidation of polysilicon films. Grain size is of interest in resistivity problems because carrier mobilities and lifetimes are affected by the need to cross grain boundaries. Doping, etching, and oxidation all occur at rates different in grain boundaries than on the exposed faces of grains at the polysilicon surface.

Texture analysis   An issue sometimes related to grain size analysis is the texture of the polysilicon film. Texture is a measure of the orientation of the grains in the film. A sample consisting of a large number of completely randomly oriented grains is said to have "no texture." If the grains in the film have some preferential orientation, for example with a larger than random part of them growing with their 111 axis perpendicular to the substrate, then they are said to have some texture, in this case 111 texture. The texture of a polysilicon film also influences resistivity, dopant in-diffusion, etching rates, and oxidation. Texture is most conveniently measure by X-ray diffraction (XRD). Figure 2.5*a* shows an XRD measurement of completely random grains from a powder sample. Figure 2.5*b* is a measurement taken from a 4000-Å-thick polysilicon film deposited on silicon dioxide grown on single-crystal silicon. The signal-to-noise in the polysilicon measurement is lower because the film is thin to the sampling X rays, and even though the sample is aligned for analysis so that the X rays strike the film at a very shallow, glancing angle (to give them a longer path length in the polysilicon film), the volume of material producing the diffracted signal is very small. The substrate is, therefore, also sampled (see the peak labeled *substrate*). Also, there is some interference between a peak from the underlying oxide and the 111 peak from the polysilicon film. Although some corrections would need to be made to these peak intensities to obtain a quantitative

                                        POLYSILICON CONDUCTORS   Chapter 2

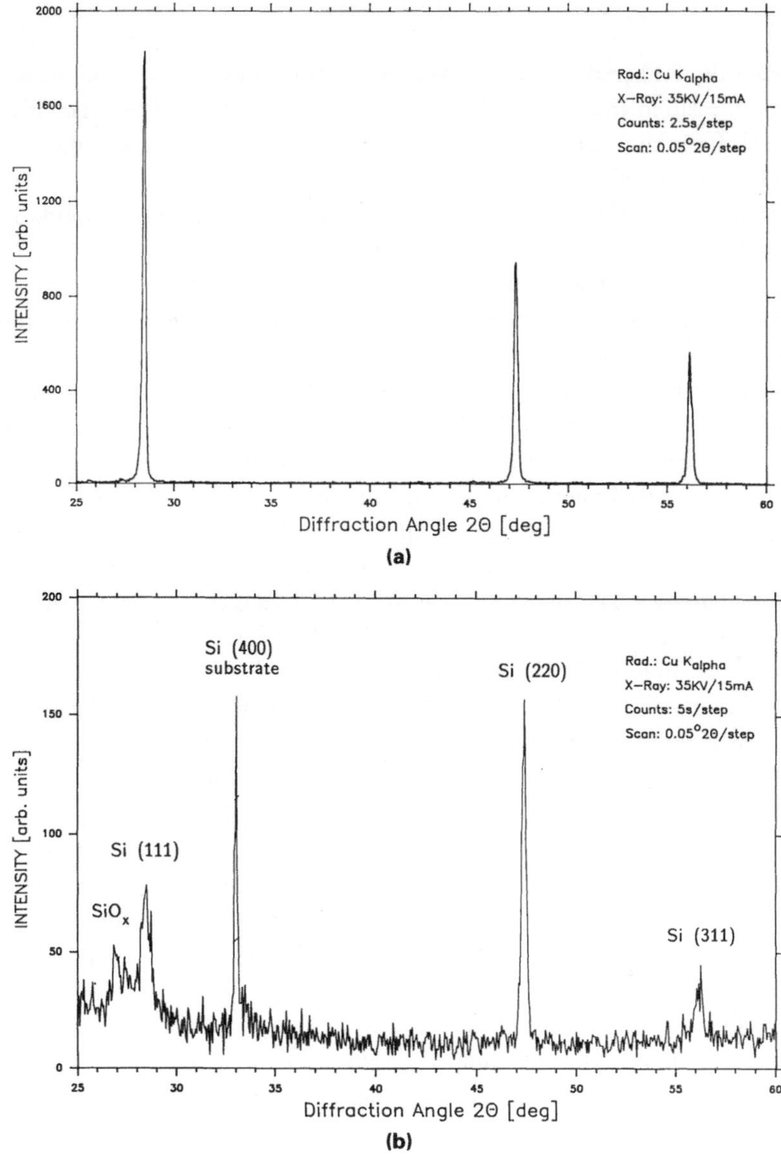

Figure 2.5 (*a*) XRD plot of NBS standard fine Si powder. This diffraction pattern is the same as that which would be produced by a completely random distribution of grain orientations in a polysilicon film. (*b*) XRD plot of 4000-Å-thick polysilicon film as-deposited on a 1000-Å-thick thermal oxide film on a Si substrate.

measure of the different grain orientations, such as a correction for the very small film thickness in this case, it is obvious from looking at this plot and comparing it to the random measurement in Figure 2.5*a* that the grain growth is not random. It is seen that, in this case, 111 oriented growth is quite weak compared to 110 (measured using the 220 peak), as is the 311 orientation. In this case, 100 growth, which must be measured from the 400 peak, cannot be determined because of the strong 400 peak from the substrate. This film has a 110 texture. Texture measurements are becoming more commonplace because of recent advances in thin-film X-ray diffractometers and the value in measuring texture as an indicator of oxidation and doping problems.[3]

When the polysilicon is deposited on single-crystal silicon, there can be a preferential epitaxial orientation in the deposited film, provided the surface is adequately clean prior to deposition. XRD texturing measurements can only show when this is not the case, because the overwhelming strength of the substrate peaks prevents measuring the epitaxially aligned grains. In this case, a comparison of randomly aligned and channeled Rutherford backscattering spectrometry (RBS) measurements can be used to give the percentage of epitaxially aligned grains by doing the channeling with the beam aligned with the substrate (see the chapter on RBS in the *Encyclopedia* and also Reference 12).

Surface morphology   The shape or form of the surface is often of concern. The amount of surface roughness, including both spatial frequency and amplitude, can be a factor in, for example, dielectric leakage between polysilicon lines in an EPROM stack or in reflectivity effects in lithography. The surface morphology can be seen in SEM or TEM cross sections, as seen in Figure 2.3. An image like this will show only a very small fraction of the surface. Mechanical profilometers which drag a stylus over the surface can map out the surface with a few nanometer height resolution, but they are destructive because contact has to be made with the surface. There are optical profilometers, but they don't have adequate lateral spatial resolution. A top- or plan-view SEM image will show contrast only when there is a significant change in surface angle. The STM will give a more quantitative representation of the surface, but again, only over a relatively small surface area (about 100 μm on a side). For rapid measurement of the surface roughness components over a large area, such as a semiconductor wafer, laser scatterometry (LS) is the best method. It will give the mean angle and amplitude of the roughness and the spatial frequency of the roughness from a subsecond measurement at each point of interest. The results can then be displayed in an isoparameter contour plot of the wafer surface. This doesn't produce a microscopically detailed image as does the STM, but it is much faster and maps out larger areas.

High-Quality Polysilicon

In polysilicon deposited for solar cell and thin-film transistor applications, there is much more interest in monitoring defects within the grains, measuring grain sizes, and mapping the corresponding grain boundaries. Defect and grain boundary mapping in these applications is frequently done using electron-beam-induced current (EBIC) and laser-beam-induced current (LBIC) measurements. These two techniques have the advantage of measuring the electrically active sites in the film. In these applications, where carrier mobility is a major concern, it is even more important

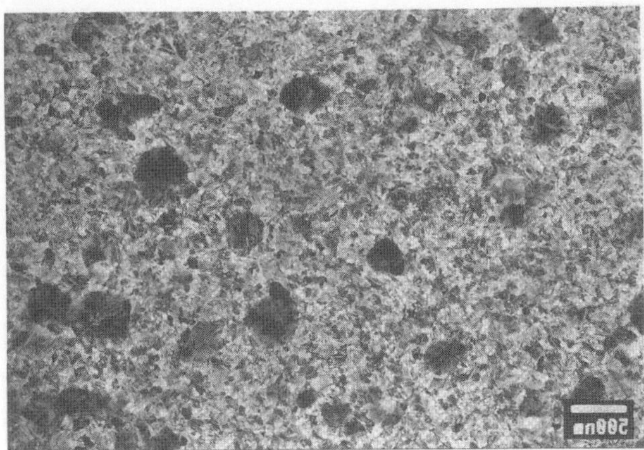

Figure 2.6   Plan-view TEM image of a "hazy" polysilicon film, as deposited on a thin thermal oxide film on crystalline Si.

to do a complete electrical characterization of the material. EBIC identifies sites where electrons generate carriers in the film, and LBIC identifies sites where light generates them. Another technique used here is photoluminescence (PL). In this technique, luminescence generated by an incident laser beam is analyzed to identify impurity states in the bandgap. A second technique used to measure impurity states in the bandgap is deep-level transient spectroscopy (DLTS). DLTS has the sensitivity to measure a very low concentration of impurity states in the bandgap, but only on high-quality polysilicon. For "average"-quality polysilicon, the states caused by crystalline defects (crystallographic impurities), both within the grains and in the grain boundaries, are in a much higher concentration than those caused by chemical impurities and overwhelm their DLTS signal. The chemical impurity states can be very important in trapping and scattering charge carriers even though they are present in concentrations that are below the detection limits of the most sensitive of the conventional analytical techniques, such as dynamic SIMS and TXRF. Although EBIC, LBIC, PL, and DLTS do not directly identify the chemical species responsible for these states, they do rely on the same charge carriers that are important in the operation of the device to detect the impurity states, and thus they have the sensitivity necessary to detect them. These techniques can then be used in combination with carefully controlled experiments to inject known impurities into the material and with the most sensitive chemical analytical techniques to identify the chemical sources of the scattering and trapping sites. With the help of these techniques, thin-film polysilicon material with mobility in excess of 100 cm^2/Vs and grain diameters in excess of 1 μm has been deposited.

Integrated Circuit Fabrication Issues

The two primary problem areas of polysilicon deposition in a semiconductor wafer fab are substrate cleaning (primarily of residues from prior processing steps) and

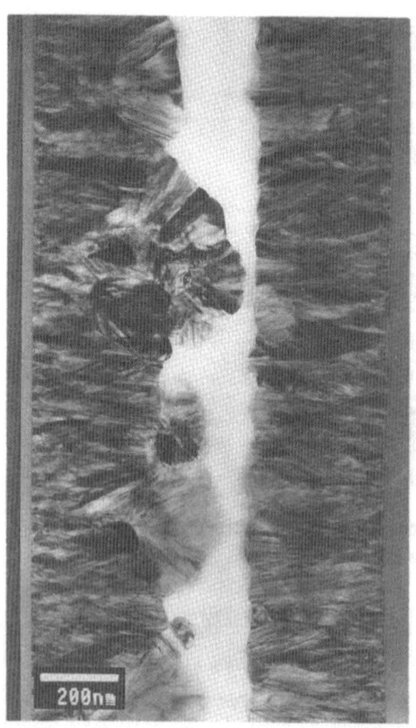

Figure 2.7   Cross-sectional TEM image of both the "hazy" and a "normal" polysilicon film. Since TEM cross sections are made by gluing samples face-to-face, it is often convenient to stack good and bad samples together for direct comparison. The "hazy" sample on the left happened to have been deposited on a thin thermal oxide film, but otherwise both samples were grown under the same conditions. The cross section cuts through a high density of the large, irregular nodules which give the sample its hazy appearance in unmagnified reflected light.

growth system control (including control of contamination from the system, system pressure, and substrate temperature). Substrate preparation, source gas purity, and growth system cleanliness are all extremely important. These are all equally easy ways to introduce contamination into the process. Contamination levels of 1 part in 10^4 may result in a substantial change in resistivity. These problems are frequently first seen as either increased surface roughness or as variations in film thickness. The roughness, if it is sufficient, may be visible to the eye as a hazy appearance instead of a shiny, mirrorlike surface on the as-grown poly.

Figure 2.6 shows a TEM plan-view image and Figure 2.7 a cross-sectional image of a layer of polysilicon with haze. The plan-view shows a relatively uniform matrix of 200–1500 Å polysilicon grains with a few large, irregular grains at a mean spacing of about 1 μm. The left side of the cross section shows the hazy polysilicon deposited on silicon dioxide which was thermally grown on a single-crystal silicon. The right side, for comparison, shows a "normal" polysilicon film which was directly deposited on single-crystal silicon. The cross section of the hazy polysilicon is through a region which contains a high density of the large, irregular grains, and

                                                        POLYSILICON CONDUCTORS   Chapter 2

shows that they are generated at various depths in the film by polysilicon grains that seem to be growing radially outward from a point. The source points are not clearly atypical in the TEM image. SIMS analysis of this hazy film showed a high level of chromium contamination of the film.

In many cases, the problem-solving process requires making many measurements to look for across-wafer variations in important materials parameters, or even wafer-to-wafer variations. This might come up, for example, in investigating source gas depletion effects in LPCVD tubes where the silicon in the source gas is depleted by deposition around the edges of the wafers before diffusion carries it to the center of the wafers, or where it is depleted by deposition on the first wafers down the tube from the source inlet before it reaches the wafers farther downstream. The techniques of choice in these cases will be those that are fast, repeatable, and require no sample destruction, or even preparation. Tools such as the four-point probe resistivity measurement systems or optical systems for measuring interferometry, ellipsometry, or scatterometry are available with provision for rapid, large-area mapping of their measured parameters. Tools such as the TEM or STM, which require much more time per analysis point, are less valuable in this type of analysis, except in final verification of causes, when one can select a limited number of samples, or regions on a sample, for such analysis.

2.3 Doping

Presently, three types of doping are used in the semiconductor industry: in situ doping (in which the dopant is introduced during the material growth process); diffusion doping (in which the dopant diffuses into the material from a gaseous or solid source during high-temperature treatment); and implanation doping (in which high-energy ions containing the dopant are directed into the material). For polysilicon, in situ doping has not been developed to a widely useful stage, diffusion doping is the most used, and implantation doping is used selectively.

Some of the important materials issues in doping are the amount of grain growth, the amount of texturing in the film after grain growth, the development of surface roughness, the dopant distribution in the polysilicon, and the incorporation of the dopant into the silicon lattice to provide electrical activation.

If thermal diffusion is the source of the dopant then grain growth, texturing and surface roughness development may all occur during the doping process. These are all best monitored by the techniques suggested in the previous section on deposition: TEM for grain-size analysis, XRD for texturing, and LS for surface roughness measurement.

Dopant Distribution

If the polysilicon film has not been in situ–doped during deposition, frequently the next step is thermal diffusion doping using $POCl_3$ as a source of phosphorus for doping the polysilicon. This process involves heating the film to approximately

900 °C in flowing $POCl_3$ plus a carrier gas. During this process, some of the phosphorus diffuses into the polysilicon and at the same time the surface of the polysilicon film becomes oxidized. The amount of phosphorus incorporated into the polysilicon and its distribution in the film is best monitored by dynamic SIMS. Dynamic SIMS involves profiling into the film while simultaneously measuring the secondary ion yields. It has a wide dynamic range and low detection limit, which may be necessary in this measurement. With the use of standards, generated by ion implantation into similar grain-sized polysilicon, dynamic SIMS has a good quantitative accuracy. Knowing the phosphorus content and distribution in the film doesn't directly give the conductivity, because some of the phosphorus can be left in interstitial positions in the silicon lattice or segregated in the grain boundaries. If four-point probe resistivity measurements show a discrepancy with the dynamic SIMS phosphorus measurements, spreading resistance profiling (SRP) is a good comparative tool that may help resolve these differences. SRP will give a depth profile, comparable to the dynamic SIMS, but one that shows the conductivity resulting from the electrically active phosphorus concentration. Differences between the SRP profile and the dynamic SIMS profile indicate unactivated dopant.

Deglaze

The oxide film produced as above is a chlorine- and phosphorus-rich SiO_2 film called a glaze. Since this glaze is a mediocre insulator, it must be removed before the next step if that step is metal or silicide deposition-which requires a good electrical contact, or the formation of a high-quality insulator which will have to withstand high electrical stress. Therefore, when the $POCl_3$ diffusion doping process is used, the next step is a deglaze in dilute HF to remove the glaze. Because of the high phosphorus, chlorine, and oxygen levels in the glaze, and because of the high spatial resolution available in AES, it is the best tool to use in verifying the effectiveness of the deglaze process. A TEM cross section of a polysilicon film with the glaze on it will show that the glaze outer surface is smooth, whereas the polysilicon surface has increased in roughness amplitude due to the grain growth that has occurred and due to the unequal oxidation rate of the different crystal faces. Thus, an insufficient HF deglaze will leave "pools" of oxide behind in the valleys of the polysilicon surface. The spatial resolution of AES is useful in showing if complete removal of the glaze has occurred.

Ion Implantation Doping

Ion implantation is often used in place of diffusion doping. It produces a different depth distribution of the dopant and extensive damage in the upper part of the polysilicon film. A TEM cross section of an ion implanted film will show the amorphization of the top portion of the polysilicon film and the remaining, untouched grain structure of the rest of the film. After implantation, the polysilicon must be annealed to recrystallize the amorphous layer and activate the dopant. Often problems arise at this point because of the dopant going into the poly grain boundaries, which results in lower than anticipated carrier concentrations (higher resistivities) and grain boundaries that oxidize rapidly. Identification of this situation requires careful analysis with a high spatial resolution technique to map the phosphorus distribution, such as dynamic SIMS with a field emission ion source, or, if the phosphorus level is high enough, EDS mapping in a TEM.

In situ doping during deposition is receiving increased attention. There are problems with in situ doping such as reduced deposition rates, dopant incorporation and distribution, impurity incorporation from the dopant source, and dopant uniformity across the wafer. All of these are amenable to analysis using dynamic SIMS, SRP, TEM, four-point probe, etc. There are no characterization problems that appear to be unique to in situ doping.

2.4 Patterning

Lithography

During the lithography part of polysilicon patterning, there is very little change in the polysilicon. Resist deposition is seldom a problem. Although resist bakes have been shown to cause grain growth in aluminum, they are not high enough to affect the polysilicon grains significantly. Exposure and development of the resist does not affect the polysilicon.

Etching

During the etch part of polysilicon patterning, there are a few problems that may occur. "Poly extras" is the term sometimes used for the appearance of small, cone-shaped islands of polysilicon residue remaining after the etching is complete. These are caused by contamination remaining in small patches on the polysilicon surface where the resist has supposedly been removed to allow the polysilicon to be etched. The contaminant is usually incompletely removed resist. It blocks the etch process locally, forming a shadow cone of unetched polysilicon beneath it. As the etch proceeds, it may eventually remove the contaminant, leaving only the cone. Thus, after etching, the poly extras may not have any remaining evidence as to the source of the problem. In that case, it is necessary to go back and look at some material from a stage before etching began. The carbon in incompletely removed resist is easily detected by AES, although the measurement is complicated by the solid resist sidewalls between which the tiny patches of remaining carbon must be detected. In addition, the resist typically causes substantial charging problems during Auger analysis. There are ways to remove the undeveloped resist to gain access to the region to be analyzed.

At times, over-etching occurs (occasionally to remove "poly extras") which can result in damage to the underlying silicon. Thermal wave (TW) analysis is often used to detect this damage. TW analysis involves only low-power laser beams impinging on the surface and thus is nondestructive. TW analysis provides a substrate damage value which, although not an absolute quantitative measure of substrate disorder, is a valuable comparative diagnostic for gauging substrate degradation (for example, during process development which may involve variation of many plasma parameters).

Although it is much more a function of etch parameter variations than of polysilicon materials properties, another problem which frequently arises is that of variations in the cross-sectional profile of the etched polysilicon lines. The standard procedure for monitoring this problem is to prepare samples destructively by cleaving etched wafers perpendicular to the poly lines, coating the samples, and then evaluating them in cross section in an SEM. The time and wasted material in this process can be significantly reduced by using LS, which has been shown to produce signatures that are extremely sensitive to dimensions and angles in periodic fields of etched structures. This technique requires only the momentary reflection of a light beam from the area of interest to produce a signature which can be compared to that of a properly etched area.

2.5 Subsequent Processing

During most of the subsequent processing steps that polysilicon layers may see, the only process input which produces any change is the time-temperature product which results in grain growth and dopant diffusion effects. The two exceptions are the deposition, anneal, and oxidation steps involved in the formation of polycide (metal silicide on polysilicon) structures and the dielectric encapsulation of the polysilicon.

Polycides

The polycide formation process will be illustrated with the example of a tungsten–polycide system. Various other metal–silicide systems are used, but many of the materials-related problems are similar. The silicide film is formed either by the deposition of a pure metal film and an anneal to form the silicide layer or by direct deposition of the metal silicide onto the polysilicon. In the tungsten–polycide case, it is usually done through the CVD of $WSi_{2.2}$. (The excess silicon will be depleted during a subsequent oxidation step.) Two important materials issues in this step are that the polysilicon surface must be clean so that the silicide layer will adhere to it and that the deposition system must have a low background pressure so that tungsten oxides are not formed during deposition. The glaze discussed previously, which is formed during some types of doping, must be removed, as must any other contaminants picked up after polysilicon deposition. For the level of surface contamination control required here, AES is typically the best tool to check cleanliness. Volume contamination in the deposited silicide, principally oxygen, picked up from poor deposition system base pressures may also be found by AES, although post-deposition profiling by dynamic SIMS may be required to see the very low levels of oxygen which can begin to form tungsten oxides. Monitoring the Si/W ratio, which can be of concern, is best done by RBS, since differential sputtering effects between the tungsten and the silicon make all sputter-profiling techniques problematic.

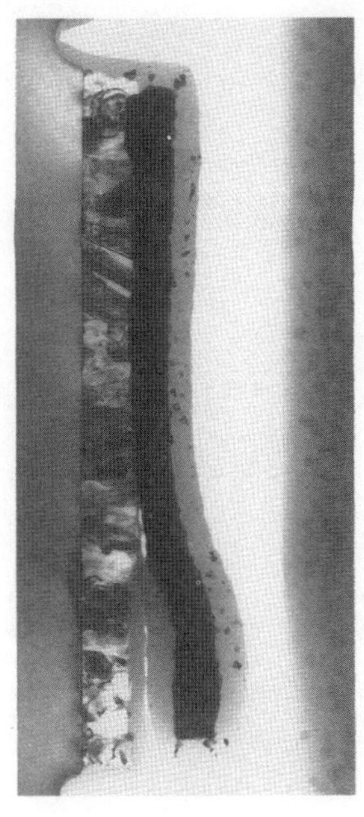

Figure 2.8 TEM cross section of a polycide line after oxidation showing, in the bottom third of the micrograph, a separation between the polysilicon (second band from the left) and the tungsten silicide (third band from the left). The oxide growth between the separated layers is also clearly seen.

Once the polycide lines are formed, the next step is to encase them in an oxide film formed by thermal oxidation. Thermal oxidation of the polycide lines provides an SiO_2 film on the exposed polysilicon and, due to the thermodynamics of the system, an SiO_2 film is also grown on the WSi_2 layer. The silicon required for SiO_2 formation on the WSi_2 comes from silicon out-diffusion from the excess silicon that was deposited in the tungsten silicide and, when that is depleted, from the underlying polysilicon layer. This is another reason the tungsten silicide–polysilicon interface needs to be clean since silicon must be free to diffuse across it. This process can be monitored by RBS, which shows the Si/W ratio decreasing to 2.0, the underlying polysilicon getting thinner, and the SiO_2 layer growing as the process proceeds.

Figure 2.8 is a TEM micrograph showing a polycide line in which the silicide layer separated from the polysilicon layer either before or during the oxidation step. There is a very high stress at this interface, so it is imperative that it be kept clean to provide good adhesion and to avoid lifting such as this. In the figure, the oxide layer can be seen covering all of the exposed (silicide and polysilicon) surfaces. The different contrasts in the TEM micrograph are caused by the differing nature of the materials: the high atomic number tungsten silicide makes it opaque (dark band) to the TEM electron beam; the polysilicon shows diffraction effects which vary along the band, giving varying contrast; and the oxide layer in the separated region causes a moderate uniform attenuation of the electron beam.

Dielectric Encapsulation

Polysilicon lines on ICs are usually covered by an SiO_2 layer to provide chemical passivation or electrical isolation or both. If only low-field electrical isolation is

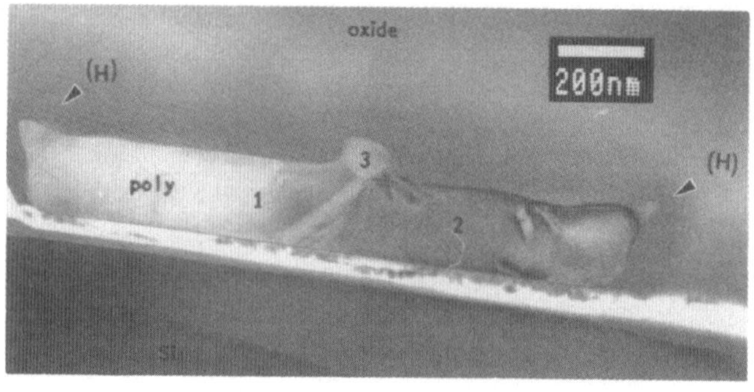

(a)

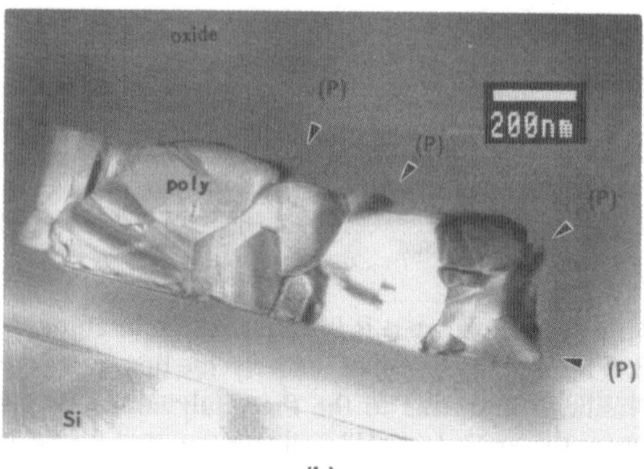

(b)

Figure 2.9 (*a*) A TEM cross section of a polysilicon line after thermal oxidation showing three distinct grains labeled 1, 2, and 3 and horns formed at the corners of the line. The bright band running across the image is a crack in the very delicate TEM sample. Grain number 3 is of an orientation such that it did not oxidize as rapidly as grains 1 and 2, since the surface of the polysilicon line was nearly flat prior to oxidation. (*b*) A TEM cross section of another polysilicon line showing the formation of protrusions at the intersections of various grain boundaries with the polysilicon surface. Horns at the original 90° corners of the line can also be faintly seen.

required, this is usually accomplished by a low-temperature CVD oxide. At times, stress can be created in this system which can be monitored by Raman spectroscopy. The Raman lines from polysilicon show a wavelength shift resulting from stress. The deposited oxides are not Raman active and thus provide no interfering signal and permit measurement through them.

When high electric field isolation is required or if the oxide is an active electrical component, like the dielectric in an MOS capacitor, thermal oxides or dielectric stacks such as thermal oxide/deposited nitride/thermal oxide are used. Thermal oxidation of polysilicon can be compared to oxidation of an array of adjoining silicon crystals (the polysilicon grains). However, certain effects can lead to problems. For example,

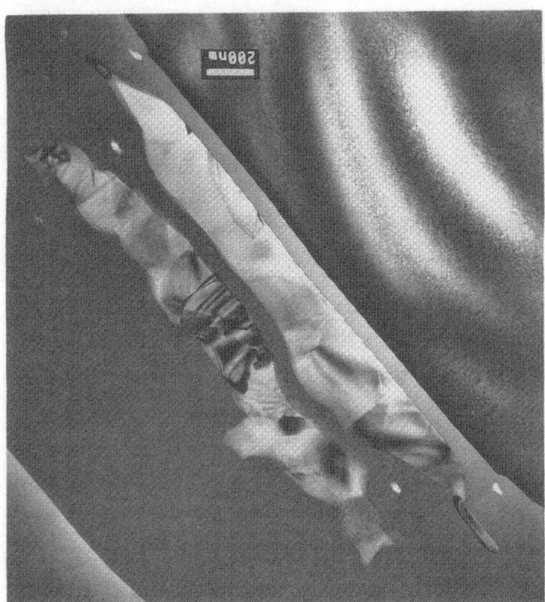

Figure 2.10   A cross-sectional TEM image of two polysilicon lines which were originally roughly rectangular in cross section, before oxidation. The oxidation has produced horns, a protruding grain, voids (bright spots) in the oxide, and detached nodules of Si floating in the oxide.

it is known that the different crystal faces of silicon oxidize at different rates. Thus even a perfectly smooth polysilicon surface, when oxidized, develops surface roughness. Grains with an exposed 111 surface will oxidize most rapidly, those with exposed 110 faces less rapidly, and those with exposed 100 faces will oxidize least rapidly. The oxidation process is further complicated by the volume expansion that occurs upon oxidation. Because of this, the presence of a corner, either present before oxidation or created during the oxidation process, results in very high stress on the silicon in the corner which, in turn, reduces the oxidation rate. Thus the oxidation rate increases with distance from the corner, causing the corner angle to become sharper as the oxidation proceeds. This leads to the creation or enhancement of surface protrusions, "horns" at corners of poly lines and, when in some cases the oxide completely pinches off an island of silicon, inclusions of silicon floating in the oxide. Examples of these anomalies can be seen in Figures 2.9*a* and 2.9*b*, which are TEM cross sections of oxidized polysilicon lines. In some extreme cases, clean, rectangular polysilicon lines can become completely distorted, as shown in Figure 2.10.

Although the TEM can produce interesting pictures illustrating these problems, it is possible to monitor the increase in surface roughness nondestructively and much more rapidly. This can be done by either SE or LS. Both techniques rely on the reflection of a light beam from the sample surface. The growing surface oxide is transparent to the light, permitting the measurement of the polysilicon roughness through the oxide.

References

1   T. Kamins. *Polycrystalline Silicon for Integrated Circuit Applications.* Kluwer, Boston, 1988.

2   "Polysilicon Thin Films and Interfaces." *Proceedings*, Vol. 182. (T. Kamins, B. Raicu, and C. V. Thompson, Eds.) Materials Research Society Symposium, Materials Research Society, Pittsburgh, 1990.

3   J. M. Drynan and T. Kikkawa. *Appl. Phys. Lett.* **58**, 610, 1991.

4   J. C. Bean. *J. Crystal Growth.* **81**, 411, 1987.

5   W. Kern and D. A. Puotinen. *RCA Rev.* **31**, 187, 1970.

6   R. S. Hockett and W. Katz. *J. Electrochem. Soc.* **136**, 3481, 1989.

7   T. Shiraiwa, N. Fujino, S. Sumita, and Y. Tanizoe. In *Semiconductor Fabrication: Technology and Metrology, ASTM STP 990.* (D. C. Gupta, Ed.) American Society for Testing and Materials, Philadelphia, 1989, p. 314.

8   Y.-W. Mo, R. Kariotis, B. S. Swartzentruber, M. B. Webb, and M. G. Lagally. *J. Vac. Sci. Technol. A.* **8**, 201, 1990.

9   I. An, H. V. Nguyen, N. V. Nguyen, and R. W. Collins. *J. Vac. Sci. Technol. B.* **9**, 632, 1992.

10 H. Watanabe, N. Aoto, S. Adachi, T. Ishijima, E. Ikawa, and K. Terada. *Appl. Phys. Lett.* **58**, 251, 1991.

11 D. E. Aspnes, W. E. Quinn, and S. Gregory. *Appl. Phys. Lett.* **56**, 2569, 1990.

12 C. K. Huang and A. Feygenson. In "Polysilicon Thin Films and Interfaces." *Proceedings*, Vol. 182. (T. Kamins, B. Raicu, and C. V. Thompson, Eds.) Materials Research Society Symposium, Materials Research Society, Pittsburgh, 1990.

3

Silicides

S. P. MURARKA

Contents

3.1 Introduction

Most of the silicide research, performed from the beginning of this century through the 1960s, used powder metallurgical techniques to produce these materials.[1–4] Silicide–related studies have primarily focused on the investigation of fundamental properties such as electrical resistivity, high-temperature stability, and corrosion resistance, as well as the silicide crystal chemistry and metal–silicon phase diagrams. The possibility of using silicides as conductors (i.e., as Schottky barriers and contacts, as well as gate and interconnect metallization) in silicon integrated circuits (ICs) motivated thin-film silicide research.[5–17] Besides the measurements of Schottky barrier heights and resistivities, the intermetallic formation in the metal–silicon systems, reaction and interdiffusion kinetics, stability at the processing and operating temperatures, oxidation and etching characteristics, and epitaxial growth on silicon have all been investigated in the past two decades.

The primary thrust of very large scale integration (VLSI) has been the miniaturization of devices to increase packing density, achieve higher speed, and consume lower power. The continued evolution of smaller devices has aroused a renewed

53

Silicides	Formed by Starting from	Sintering Temperature, °C	Resultant Resistivity, $\mu\Omega\cdot$cm
$TiSi_2$	Metal on polysilicon	900	13–16
	Cosputtered	900	25
$ZrSi_2$	Metal on polysilicon	900	35–40
$HfSi_2$	Metal on polysilicon	900	45–50
	Cosputtered alloy		60–70
VSi_2	Metal on polysilicon	900	50–55
$NbSi_2$	Metal on polysilicon	900	50
$TaSi_2$	Metal on polysilicon	1000	35–55
	Cosputtered alloys	1000	50–55
$CrSi_2$	Metal on polysilicon	700	~600
$MoSi_2$	Metal on polysilicon	1100	~90
	Cosputtered alloy	1000	~100
WSi_2	Cosputtered alloy	1000	~70
$FeSi_2$	Metal on polysilicon	700	>1000
$CoSi_2$	Metal on polysilicon	900	10–18
	Cosputtered alloy	900	25
$NiSi_2$	Metal on polysilicon	900	~50
	Cosputtered alloy	900	50–60
PtSi	Metal on polysilicon	600–800	28–35
	Cosputtered alloy		
Pd_2Si	Metal on polysilicon	400	30–35

Table 3.1   Resistivities of various silicides.

interest in the development of new metallization schemes for low-resistivity gates, interconnections, and ohmic contacts. This interest in new metallization has been stirred by the fact that with the scaling down of the device sizes the linewidth gets narrower and the sheet resistance contribution to the resistance and capacitance (RC) delay increases. The characteristics of metal oxide semiconductor (MOS) devices depend on several parameters of which the RC time constant is most important. The higher the RC value, the slower the speed of an operating device. The R and C represent the effective total resistance and capacitance at the gate and interconnection level. The relationship of this RC factor to device scaling is complicated, especially because of the dependence of the total capacitance on feature size.

With polysilicon sheet resistance of 30–60 $\Omega/\square$, it became apparent that the advantage of further scaling will be offset by the interconnect resistance at the gate

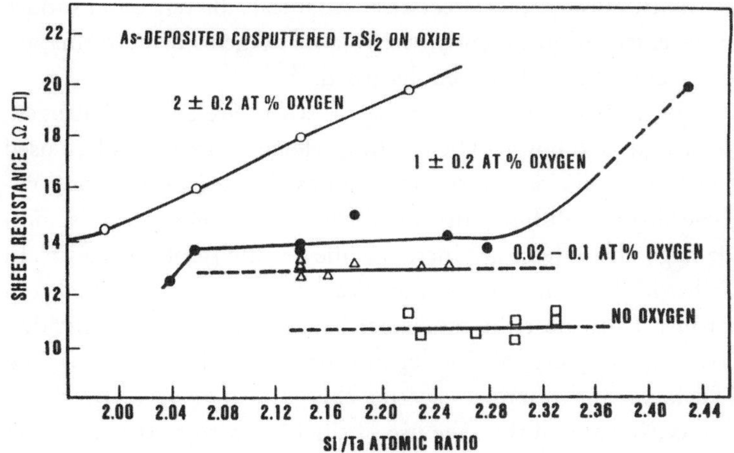

Figure 3.1 Variation in as-deposited sheet resistance of cosputtered Ta–Si deposits on oxide monitors as a function of Si/Ta ratio for several oxygen concentration ranges.

level. Most metals do not qualify for one reason or another as direct replacement of the polysilicon. Notable among the metals proposed for gate and interconnect metallization have been aluminum,[18] tungsten,[19, 20] and molybdenum.[19, 20] Use of aluminum, however, requires all postgate processing of the devices to be limited to very low temperatures, preferably below 500 °C. Use of the refractory metals tungsten and molybdenum requires complete passivation of these metals from oxidizing ambients, deposition by such means that will not lead to unwanted traps in the gate oxide, and reliable etching of the metals for pattern generation. The uncertainties associated with the stability of these metal films led to the use of silicides which attracted attention because of their low and metal-like resistivities and their high-temperature stability.

Resistivity of silicides is the single most important factor in considering them for use as metallization in the ICs. Table 3.1 lists the resistivities of silicon-rich silicides for the refractory and near-noble metals. A range of resistivities is given in each case, as reported by various workers. The variation in the resistivities is the result of variability in preparation of the silicides. The variability is caused by purity (especially oxygen contamination) and crystallinity. Oxygen increases the resistivity significantly. As an example, Figure 3.1 shows the variation of the as-deposited sheet resistance of Ta–Si deposits on oxide as a function of the Si/Ta ratio for several oxygen concentration ranges. Sheet resistance is related to resistivity ρ by the relation $R_s = \rho/d$, where d is the film thickness which was 2500 ± 100 Å for films shown in Figure 3.1. These results have been explained on the basis of the presence of oxygen and the other impurity, silicon, in as-deposited films.[21] Upon annealing, crystallization sets in and the resistance drops significantly to a value expected of a TaSi$_2$ film. For crystallized films, a small concentration of oxygen does not effectively contribute to the resistivity. However, its effect on mechanical adherence and surface roughness has been observed. More recently, the role of oxygen on the formation behavior of refractory metal silicides has been discussed.[22]

Other impurities which effectively increase the resistivity are silicon, nitrogen, carbon, phosphorus, arsenic, and boron. The last three elements are most often used as dopants in silicon. When a silicide is forming on a heavily doped silicon or polysilicon substrate, these dopants diffuse into the silicide and change its resistance. Ion implantation damage also effectively increases the resistivity of silicides.[23] In most applications, silicides films are thin and in the range of a few hundred to a few thousand angstroms. Except for the molecular beam epitaxy (MBE) silicides, the films are polycrystalline and crystal size depends on the method of formation. Grain boundaries, surfaces, and interfaces scatter charge carriers very effectively, increasing the resistance. In spite of this, the use of silicides, with resistivity values about one tenth (or lower) than those of polysilicon, improves the speed of ICs. Formation of silicides directly on the polysilicon, thus preserving the basic polysilicon MOS gate while decreasing the resistance, led to the use of silicides for gate and interconnect metallization.

Scaling down in size also means reduced junction depths. This could lead to contact problems. In particular, shallow junctions limit the use of aluminum due to its known penetration in silicon. Forming silicides in the contact windows by reaction between the silicon substrate and a thin metal layer offers the possibility of forming contacts with lower contact resistances while preserving shallow junctions.

When a metal is deposited or a silicide is formed on a semiconductor (i.e., silicon), a potential barrier to charge transfer results because the Fermi levels are required to match up. Such a barrier, commonly referred to as a Schottky barrier, arises from the difference in the work functions of the metal and the semiconductor. The height Φ_B of this barrier predicts the current flow characteristics of the metal on semiconductor device.[24] The metal–semiconductor contact resistance is a function of the height of the Schottky barrier and doping density (and type) in the semiconductor. The question arises as to what controls the barrier heights of various silicides on silicon. Careful and accurate measurement[25, 26] of Schottky barrier heights have shown that the barrier height is a property of the given metal on silicon system and is independent of the silicide phase formed. Further experiments carried out in ultra-high vacuum systems indicate that the barrier height is pinned as soon as a monolayer of the metal is deposited on a clean silicon surface.[25–27] These results, together with cross-sectional transmission electron microscopic results showing a silicon–metal abrupt interface (to within 3 Å, the resolution limit of these cross sections), clearly support the dependence of the Schottky barrier height on the metal electronic properties. In some cases, the barrier height of the metal is found to be different from that of its silicide. The difference is attributed to the diffusion of metal atoms, which create traps for electrons and holes and thus participate in the current-carrying process, into the semiconductor.[28]

Of all the silicides, only silicides of groups IVA, VA, and VIA, called the refractory metal silicides, and of group VIII, called the near-noble-metal silicides, are of particular interest to us. Fortunately, silicon-rich silicides are most stable and generally most conducting. The refractory metal silicides (with higher melting points) would be suitable for high-temperature applications. On the other hand, group VIII metal silicides will be useful for low-temperature applications.

Refractory disilicides of Mo, Ta, Ti, and W with respective thin-film resistivities of about 100–120, 50–65, 15–25, and 60–70 $\mu\Omega \cdot$cm are being used on top of the polycrystalline silicon film to reduce the resistance in large-density memories, such as dynamic random access memories (DRAMs), and in newly developed microprocessors. They were chosen over the refractory metals because of their (1) high-temperature stability, (2) oxidizability, and, most importantly, (3) retrofitability in an existing process. Their use led to the reduction of the resistance at the gate level by a factor of 5–10. More recently, with the lowering of the process temperatures to ≤900 °C, the disilicide of cobalt[29] has also become an important contender as a self-aligned silicide for application both at the gate and contact levels. PtSi can also be seriously considered for application as the contact metal on shallow junctions provided postsilicide temperatures are below 700 °C. The main reason for selecting PtSi (over $CoSi_2$) and $TiSi_2$ as contacts is related to consumption of silicon, which is significantly lower during PtSi formation.

The usefulness of the silicide metallization scheme depends upon achievement of a low resistivity, the ease with which the silicides can be formed and patterned, and the stability of the silicides throughout device processing and during actual device usage. A great number of analytical techniques are employed to follow the formation and to characterize the silicides. In this review, various characterization tools used to characterize the silicides and thus to aid in determining their properties, formation (reaction) kinetics, processing behavior, and stability are discussed. However, there is no discussion of the fundamentals and limitations of these techniques; instead, references are cited to guide the reader to the necessary information.

3.2 Formation of Silicides

Silicides are formed by annealing metal on silicon or polycrystalline silicon (polysilicon) substrates or by codepositing[30] a metal–silicon mixture (possibly a silicide when deposition temperatures are high followed by annealing at high temperatures. The kinetics of the silicide growth is generally followed by use of (1) sheet resistance, (2) Rutherford backscattering (RBS), (3) X-ray diffraction (XRD) measurements, or (4) some combination of these. More recently, ellipsometry has been used where very thin silicides (<100 nm) have been formed.[31]

Sheet Resistance Measurements

Silicides have metal-like resistivities and, therefore, measuring the resistivity ρ or the resistance of a given conducting element of the silicide provides an excellent way of characterizing the silicide. The most commonly used method to measure ρ is a four-point method which directly measures sheet-resistance R_s as defined earlier. R_s multiplied by the film thickness yields the resistivity. For known silicide resistivity, one can inversely use R_s to determine the thickness of the silicide.

Sheet resistance (R_s) measurements have extensively been used in studying silicide formation. For example, we illustrate silicide formation in the cobalt–silicon system. There are three known phases of cobalt silicides that form when a thin-cobalt film is reacted with silicon: Co_2Si, $CoSi$, and $CoSi_2$, with respective resistivities of ~60, ~150, and ~15 $\mu\Omega$·cm. In the typical Co–Si interaction, Co_2Si forms first, followed by $CoSi$. These two phases grow until all the Co is consumed, after which $CoSi$ grows at the expense of Co_2Si. The film transforms completely to $CoSi$, and on annealing above 500 °C, $CoSi_2$ forms. If the phases can be considered as parallel layers, the overall sheet resistance can be described by

$$\frac{1}{R_s} = \frac{1}{R_{s\,Co}} + \frac{1}{R_{s\,Co_2Si}} + \frac{1}{R_{s\,CoSi}} + \frac{1}{R_{s\,CoSi_2}} \tag{3.1}$$

or

$$\frac{1}{R_s} = \frac{t_{Co}}{\rho_{Co}} + \frac{t_{Co_2Si}}{\rho_{Co_2Si}} + \frac{t_{CoSi}}{\rho_{CoSi}} + \frac{t_{CoSi_2}}{\rho_{CoSi_2}} \tag{3.2}$$

Initially, R_s is determined by the first term since the cobalt is yet to react. Once cobalt reacts with silicon, the second and then the third terms begin to dominate, until, with the consumption of Co and Co_2Si, the third term determines the sheet resistance. On the formation of $CoSi_2$, the final term influences the R_s value, until only $CoSi_2$ exists and R_s is determined by the final term alone. Sitaram[32] has demonstrated the simplicity of this method by modeling the sheet resistance as a function of the process sequence leading to a combination of Co, Co_2Si, or $CoSi_2$ existing in the annealed metallic layer over silicon. Figure 3.2*a* shows a computed R_s versus reaction sequence for a 1000-Å cobalt film completely reacted to form Co_2Si, which then completely formed $CoSi$ and, finally, $CoSi_2$. Other sequences can be modeled in a similar manner. Such calculations support the experimentally observed R_s versus temperature plots[16] and show the ease with which the Co–Si reaction can be monitored, with complete formation of $CoSi$ indicated by the maximum in the plot (Figure 3.2*b*). Applebaum et al.[33] have successfully used this technique in following the reaction kinetics of the CoSi transformation into $CoSi_2$, where only the last two forms in the right side of Equation 3.2 are retained. A substitution of thickness, densities, and $CoSi/CoSi_2$ density ratio leads to the equation

$$d^{CoSi_2} = \left[\frac{1}{R_s} - \frac{d_0^{CoSi}}{\rho^{Co}}\right]\left[\frac{\rho^{CoSi_2}\rho^{CoSi}}{\rho^{CoSi} - (2.02/3.52)\rho^{CoSi_2}}\right] \tag{3.3}$$

From known initial cobalt thickness and resistivities of the two phases, Appelbaum et al.[33] were able to follow the reaction kinetics easily.

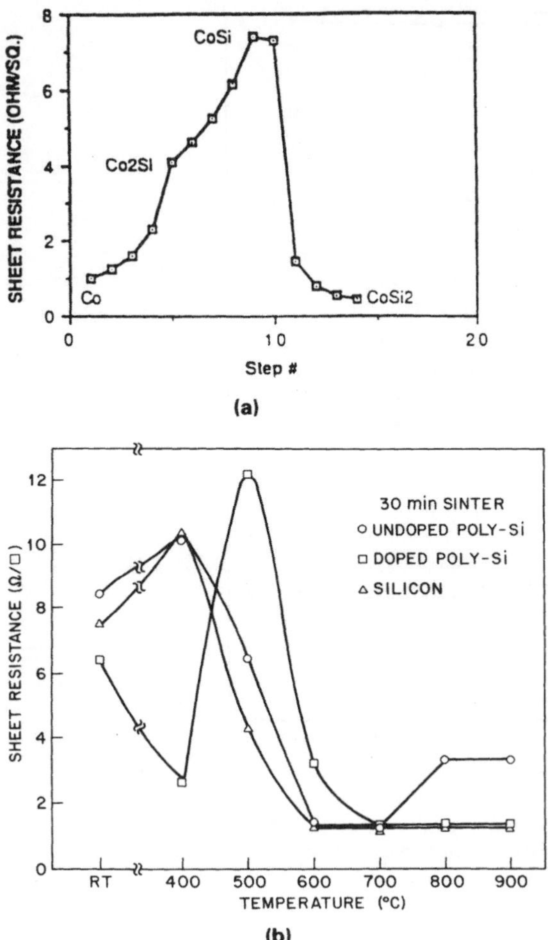

Figure 3.2 (*a*) Computed sheet resistance versus assumed reaction sequence for a 1000-Å cobalt on silicon completely reacted to form Co_2Si, then CoSi, and finally $CoSi_2$ (from Reference 38); and (*b*) experimentally obtained sheet resistance of Co on undoped and doped polysilicon as a function of annealing temperature.

The sheet resistance method, however, is based on the assumptions that (1) the substrate does not contribute to R_s and that only phases under consideration are present in the film, (2) the phases are in a parallel, layered sandwich structure, and (3) the resistivities of the phases are uncharged.

Sheet resistance measurements have also been extensively used to evaluate diffusion barriers. Figure 3.3 shows the effectiveness of 1000-Å-thick Ta_2N as the diffusion carrier between Pd_2Si and Al at least up to 1 h at 555 °C.[34] These results were confirmed with RBS, X-ray, and Auger analyses.

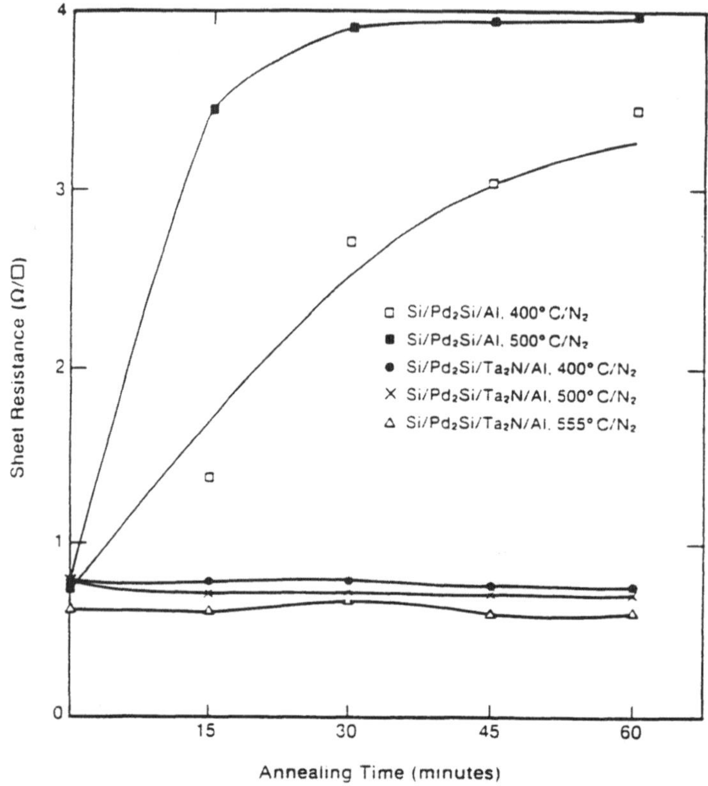

Figure 3.3 **Sheet resistance versus annealing time at various temperatures for aluminum on Pd$_2$Si with/without Ta$_2$N diffusion barrier, demonstrating the effectiveness of this barrier.**

Rutherford Backscattering Measurements

RBS[35] has been extremely popular in studying various metal–silicon reactions leading to silicide formation, especially because of its quantitative and nondestructive nature and the ease with which it can be performed. More recently, channeling RBS has been extremely useful in studying various aspects of structural analysis of solids. The technique, however, is limited to heavy elements and is therefore not suitable for B and P redistribution studies or for studies in which carbon, nitrogen, oxygen, or other light elements have to be detected and concentrations determined. Also, the technique is unable to provide chemical information, that is, phases of silicides or other intermetallics for which X-ray or electron diffraction techniques will be complementary.

Codeposition of silicides is a preferred way of forming silicides where the complete substrate–deposit system can not be subjected to high-temperature anneals or where the availability of silicon, necessary to form the silicide by reaction with metal, is limited or none. Examples are the formation of refractory silicides and

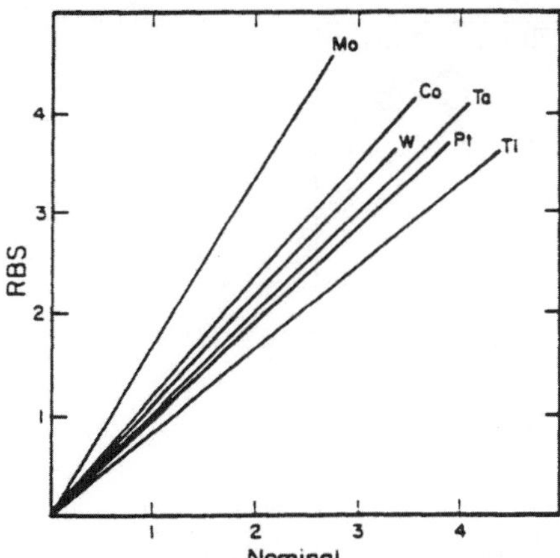

Figure 3.4   **Calibration curves for various Si/M ratios in cosputtered metal silicide films, ratios determined by RBS versus nominal ratios determined from sputtering parameters.**

deposition of the silicides on SiO_2 or on a very shallow n–p or p–n junction. Codeposition can be carried out by sputttering, evaporation, and chemical vapor deposition (CVD) processes. Deposition must be controlled to yield the desired silicon to metal (Si/M) atomic ratio in the film. RBS is used to monitor such depositions and thus provide Si/M control in the films. Experience has shown that sputtering provides the best control and flexibility in adjusting the Si/M ratio which is controlled in the equipment by calibrating the individual silicon and metal target sputtering rate versus applied target power. However, the actual Si/M ratio in the deposited film is invariably different than the nominal Si/M ratio based on the previously mentioned calibration. Actual Si/M ratios could be easily determined by use of RBS. Figure 3.4 shows the plots of actual Si/M ratios as a function of the nominal Si/M ratios for Ti–Si, Mo–Si, Co–Si, Pt–Si, Ta–Si, and W–Si.[30] It must be emphasized that this set of curves was obtained in a two-target dc magnetron sputtering chamber where the silicon target was in place all the time and the other, metal, target was changed as desired. The sputtering gas was argon and the pressure was 4 mtorr. A uniform distance was maintained between the substrate-holding planetaries and the targets. Target power, however, was varied according to the metal–silicon system desired. It was found that when the pressure, the target–substrate distance, or the target powers (for the same Si/M ratio) were changed, the calibration curves of the type shown in the Figure 3.4 changed. The scattering of the sputtered species and their mean free paths were affected by the above parameters. This effected a change in the Si/M ratio, and the change was different for different metal–silicon systems. The mutual scattering of silicon and metal species, and the wafer position with respect to the targets, also affected the ratios. The sputtering gas impurities—especially the moisture and oxygen—significantly affected the Si/M ratios. The shifts in the Si/M ratios, due to oxygen and moisture,

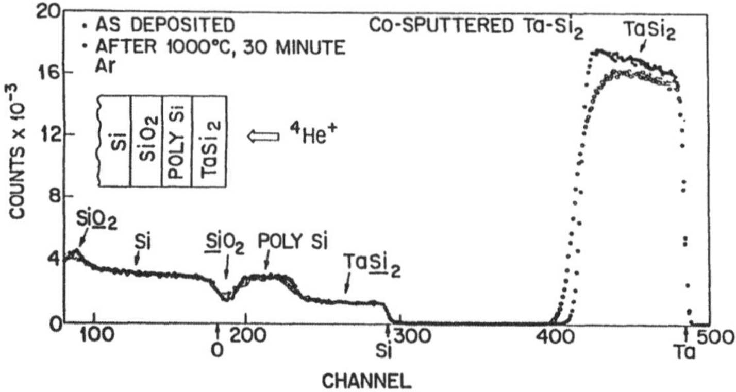

Figure 3.5 Comparision of RBS spectra of 2500-Å cosputtered $TaSi_2$/3000-Å polysilicon/1000-A thermal SiO_2/Si before and after 1000 °C/30-min anneal.

were different for different metal–silicon films. It may be pointed out that the linearity of the relationship shown in Figure 3.4 invariably fails at higher Si/M ratios not shown in this figure.

Figure 3.4 presents the calibration for the as-deposited films. Upon annealing, redistribution of atomic species occur. The redistribution is a function of the starting Si/M ratio, annealing temperature, time, ambient, the substrate, and the metal type. Typical RBS spectra of Si–Ta deposits before and after anneal are shown in Figures 3.5–3.8. In Figure 3.5, the RBS spectra of a 2500-Å cosputtered Ta–Si film (with Si/M ratio of 2) on polysilicon are displayed. Analysis of these spectra reveals that annealing led to a change in the Si/Ta ratio of the film; the ratio changed from 2.0 to 2.2. Also, approximately 460 Å of polysilicon has been consumed, and tantalum tail into the remaining polysilicon extends for about 370 Å.

Figure 3.6 shows the RBS spectra of a $TaSi_{1.3}$ film on polysilicon before and after anneal at 1000 °C.[36] This film reacts with the underlying polysilicon to produce a thicker $TaSi_{2.1}$ film. Figures 3.5 and 3.6 clearly demonstrate that there is a tendency in silicon-deficient films to absorb silicon in excess of the stoichiometric Si/Ta ratio of 2 and form a more thermally stable film with a Si/Ta ratio between 2.1 and 2.3. This conclusion is supported by the fact that the RBS spectrum of a film with a Si/Ta ratio of 2.2 is identical to one of the same film after a 1000 °C anneal in an inert ambient. These observations and those discussed in the following paragraphs led to the use of $TaSi_{2.2}$—$TaSi_{2.3}$ on polysilicon as the gate—and interconnection metallization in VLSI circuits.

Figure 3.7 shows the results of annealing a film with a Si/Ta ratio of 3.2.[36] Upon annealing for 1 h at 1000 °C, the film has rejected silicon and converted to a film with a Si/Ta ratio of 2.3. The underlying polysilicon thickness has increased by about 700 Å, and there is a pileup of about 200 Å of silicon at the silicide–cap oxide interface. Thus, in this case, silicon is diffusing out from both silicide surfaces. This

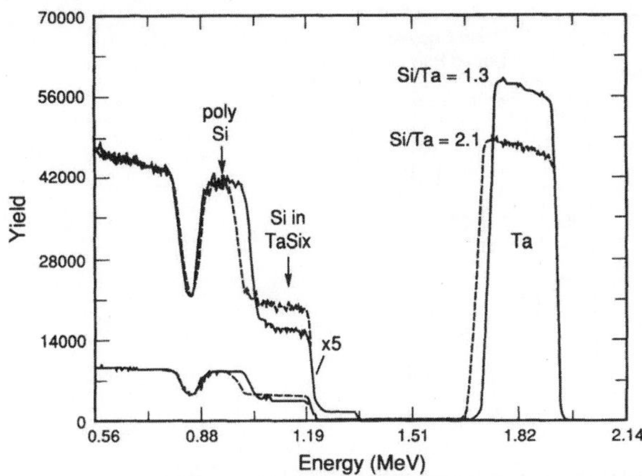

Figure 3.6 RBS spectra of the as-deposited film with a Si/Ta ratio of 1.3 and of the same film after the 1000 °C, 1-h anneal.

silicon out-diffusion depends on the anneal temperature. At lower temperatures, only silicon from the near surface volumes is able to diffuse out, making these regions rich in tantalum, as shown in Figure 3.8. This figure shows a series of tantalum RBS curves of this material (film with a Si/Ta ratio of 3.2 and capped with SiO_2) as a function of annealing temperature. In this figure, "interface" refers to the polysilicon–silicide interface and "surface" is the silicide–cap oxide interface. At intermediate temperatures silicon rejection is not complete, leaving the central volume at the original composition. Only after a 1-h anneal at 1000 °C is the film nearly homogenous in composition.

A large number of experiments with and without cap oxides have shown that in the absence of the cap oxide there is more silicon loss, apparently to the ambient.

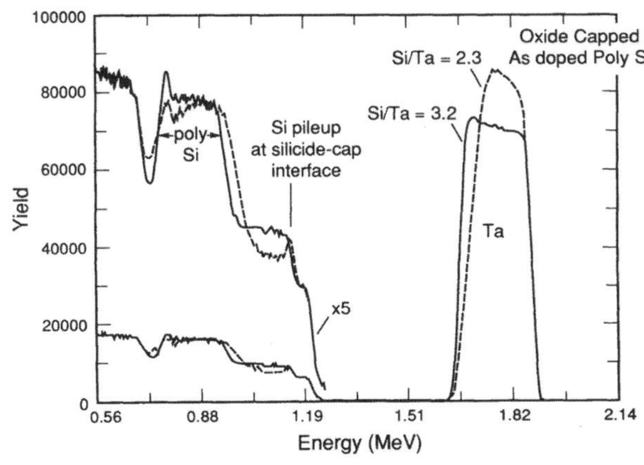

Figure 3.7 RBS spectra of the as-deposited film with a Si/Ta ratio of 3.2 and of the same film after 1000 °C, 1-h anneal.

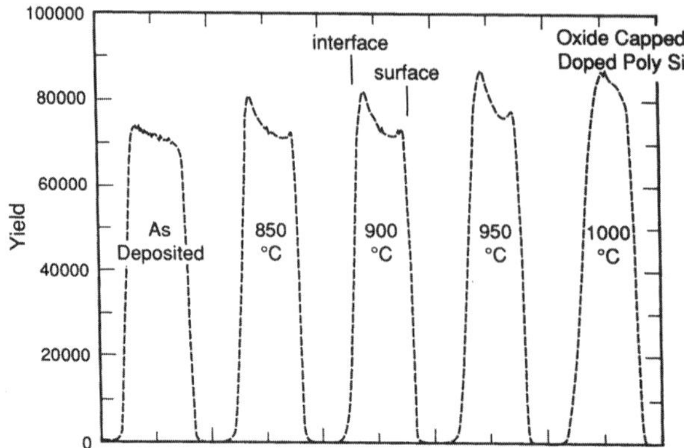

Figure 3.8   Ta peak of the film (starting as-deposited ratio of 3.2) as a function of anneal temperature. Annealing was performed after a cap plasma oxide was deposited on surface.

This leads to a significantly more tantalum-rich surface layer compared to that of a capped sample. In the latter case, the oxide on the surface prevents loss of Si to the ambient and silicon piles up at the interface, inhibiting further migration of silicon from the silicide to this interface. Rejection to the polysilicon–silicide interface,

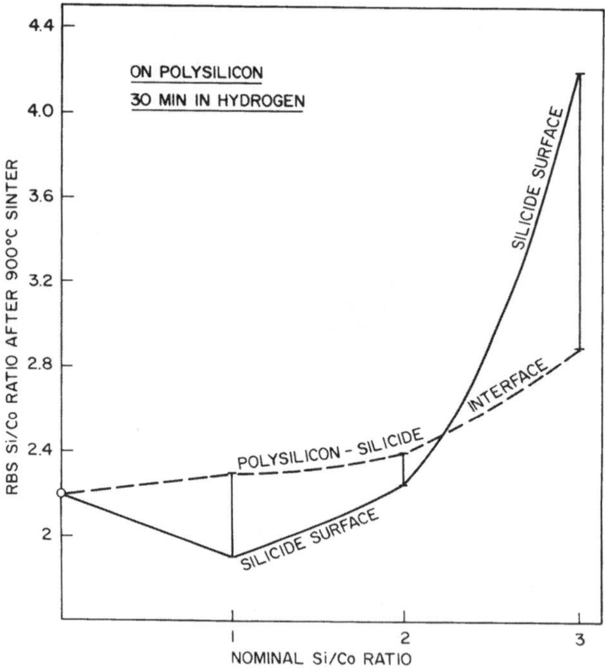

Figure 3.9   RBS (Si/Co) ratio at the silicide surface and the polysilicon–silicide interface (after 900 °C anneal) as a function of the nominal (Si/Co) ratio of the films deposited on polysilicon.

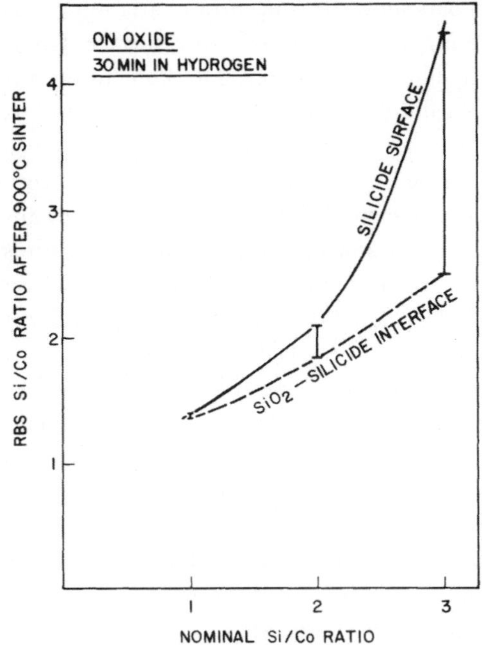

Figure 3.10 RBS (Si/Co) ratio at the silicide surface and the SiO$_2$–silicide interface (after 900 °C anneal) as a function of the nominal (Si/Co) ratio of the films deposited on SiO$_2$.

however, continues until larger pileups occur, apparently due to the crystallization of rejected silicon on polysilicon grains leading to lowering in the free-energy of the system. The higher the temperature, the greater is the out-diffusion from the film. At high enough temperatures or long enough times, silicon out-diffusion is more or less complete, leading to nearly uniformly constituted TaSi$_{2.3}$ films for capped samples and TaSi$_{2.2}$ films for uncapped samples.

Similar studies with films of Si–Ti, Si–Mo, and Si–W showed that the films with Si/Ti, Si/Mo, and Si/W ratios of 2.4–2.5, 2.1–2.2, and 2.1–2.2, respectively, were stable at annealing temperatures up to 1100 °C. In the case of Si–Co films, however, the conclusion was different.

Silicon–cobalt films with nominal Si/Co ratios of 0 (i.e., pure cobalt), 1, 2, and 3 (see Figure 3.4 for actual RBS Si/Co ratios) on polysilicon were annealed at various temperatures and RBS spectra generated. As-deposited films showed homogenous mixtures of silicon and cobalt in cosputtered films. Upon annealing, compositional inhomogenuity across the film thickness (similar to that observed for TaSi$_{3.2}$ films) developed. The extent of this compositional variation was different for different films and was affected by the nature of the substrate (polysilicon versus SiO$_2$). RBS was the most suitable technique to study these variations in compositions quantitatively. Figures 3.9 and 3.10 show the RBS Si/Co ratios after annealing at the polysilicon and SiO$_2$–silicide interface and the silicide surface as a function of nominal Si/Co ratio.[37] Vertical bars show the concentrate variation across the thickness of the silicide film from the inner interface to the top surface. $CoSi_2$ films formed by the reaction of cobalt with silicon were stable. This conclusion was extremely valuable in determining the applicability of $CoSi_2$ in VLSI circuits and its preferred formation technique.

In the case of Si–Pt deposits and PtSi formed by the reaction of Pt with silicon, the RBS studies resulted in conclusions comppletely different from those made for Si–Co deposits. Figure 3.11 shows the RBS spectra of Pt, SiPt, and Si_2Pt (i.e., films with nominal Si/Pt ratios of 0, 1, and 2 in the as-deposited state, respectively) on silicon before and after anneals at 600, 700, 800, and 900 °C.[38] Only the significant parts of the spectra are shown and compared. The Pt peak is shown in each case. As is evident, Pt reacts with silicon to form PtSi, which is unstable at higher temperatures because Pt diffuses from PtSi into the underlying silicon. The higher the temperature, the deeper is the Pt penetration in the silicon substrate. Cosputtered SiPt film behaves similarly, although the penetration and intermixing depths in silicon are relatively smaller. As far as high-temperature stability is concerned, the Si_2Pt film is best of all. There is no change in the spectrum up to 800 °C, and a comparatively small Pt-penetration in silicon is seen at 900 °C. Note that no compositional variation across the thickness of the film was detected in Si–Pt codeposits.

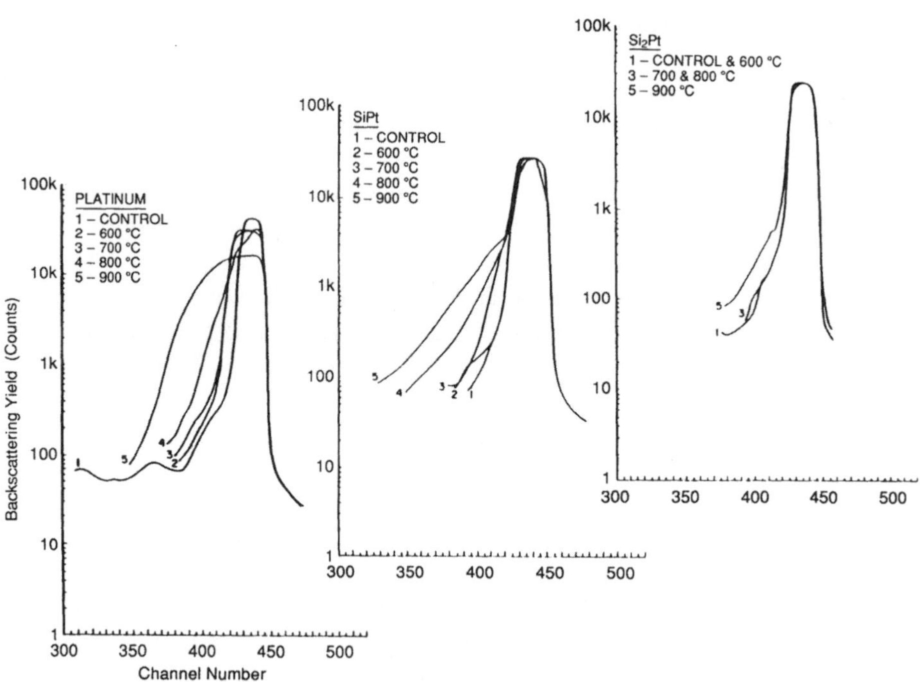

Figure 3.11 RBS spectra of Pt SiPt, and Si_2Pt films on Si after anneals at different temperatures, Only the most revealing parts of the spectra are shown.

Figure 3.12 shows the RBS calculated Si/Ti ratio as a function of the nominal as-sputtered Si/Ti ratio before and after 900 °C anneal.[39] These films were deposited on polysilicon films. Films with RBS Si/Ti ratios of less than 2.5 dissolved silicon, upon annealing, leading to a ratio of 2 or more. Films with a Si/Ti ratio of more than 2.5 rejected silicon. Films with a ratio of 2.5 were most stable during this heat treatment. These results are similar to those of Si–Ta codeposits with one difference: No compositional variation was observed in silicon-rich films. In this regard, Pt- and Ti-containing films behaved in a similar manner.

These results clearly established the usefulness of the RBS technique in demonstrating the applicability of cosputtered silicides in VLSI. It was employed as a quantitative analytical tool to elucidate the Si/M ratio and compositional instabilities, the role of oxygen and Si/M ratio in determining film resistivity, and the effect of high-temperature treatments, including oxidation, phosphorous-glass flow, and aluminum anneals. Refractory silicides of $TaSi_{2.2}$, $TiSi_{2.5}$, $WSi_{2.2}$, and $MoSi_{2.2}$ were found to be stable under these conditions. On the other hand, PtSi and $CoSi_2$ were stable only at processing temperatures lower than 700 and 950 °C, respectively.

It must be noted that RBS was also useful in establishing the usefulness of single-silicide targets in the industry. It was demonstrated that the film composition was locked to that of the target prepared by powder metallurgical techniques. There was very little variation in the film composition due to sputtering variables. For day-to-day operations, use of a single-silicide target of the appropriate Si/M ratio was adopted.

In the above examples, codeposition by sputtering was discussed. Similar studies were carried out in various laboratories around the world using coevaporated and CVD techniques. RBS was the most direct tool for determining the silicide composition and uniformity across the thickness.

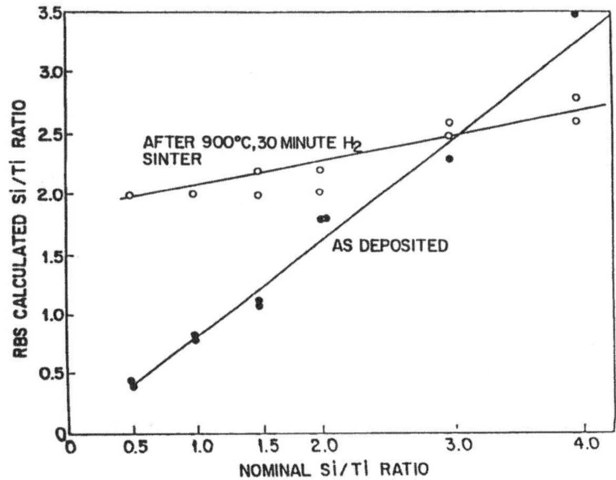

Figure 3.12 RBS calculated (Si/Ti) ratio before and after a 900 °C, 30-min, H_2 anneal as a function of the nominal cosputtered (Si/Ti) ratio.

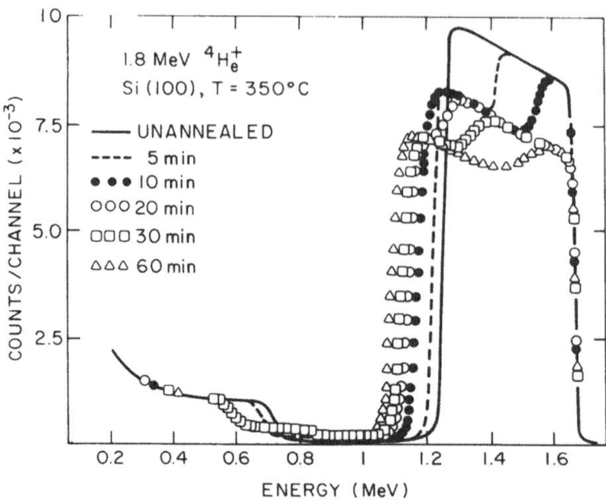

Figure 3.13   RBS spectra of samples at 350 °C for different times. The as-deposited Pt is 2700-Å thick. (From Reference 40.)

The silicides are also formed by the reaction between the metal film and the silicon substrates. RBS has been instrumental in following the reaction kinetics, determining the Si/M ratio in the phases thus formed, demonstrating the role of impurities and excess metal versus excess silicon in establishing the reaction kinetics, stability of the reactively formed silicides, and epitaxy of some silicides. Most of the initial studies of silicide formation used Pd and Pt. Figure 3.13 shows the RBS spectra of platinum on silicon as a function of the annealing time.[40] The spectra obtained from samples treated at 350 °C for different lengths of time are superimposed on each other for easy comparison. Platinum is consumed as Pt_2Si forms. In a 20-min sample, all the platinum is converted to Pt_2Si and a small amount of PtSi is found at the silicon–Pt_2Si interface. After a 60-min anneal, nearly two-thirds of the Pt_2Si film is converted into PtSi. Using such data, one can study the time and temperature dependence of the metal–silicon interaction.

RBS techniques were employed in determining the high-temperature stability of thick and thin reactively formed $CoSi_2$.[41] Stability is found to be different in reducing (forming gas) and oxidizing ($O_2 + N_2$ mixture) annealing ambients. Figure 3.14 shows the superimposed RBS spectra of Co on polysilicon composites after different temperature anneals in H_2. From the theoretical densities of Co, Si, and $CoSi_2$, 1500 Å of Co should convert to 3000 Å of CoSi and 5200 Å of $CoSi_2$. These thicknesses are consistent with the observed width of the Co signal in the as-deposited, 600 °C, and 800 °C spectra. The bump at the low-energy end of the Co peak in the 800 °C trace arises from the Superposition of the high energy portion of the Si peak (Si at the surface) with the low-energy tail of the Co signal. A similar slight overlap is also detectable in the 600 °C trace.

                                                SILICIDES   Chapter 3

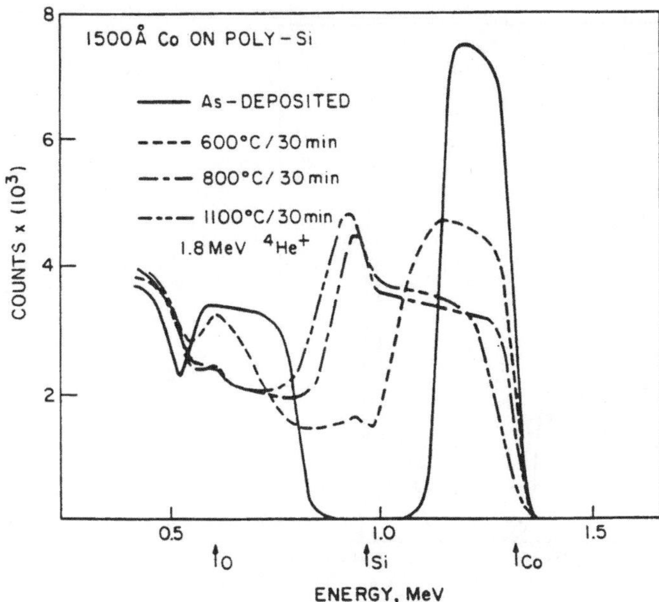

Figure 3.14   Superimposed RBS spectra of 1500-Å Co on polysilicon, as-deposited, and after 30 min forming gas anneals at 600, 800, and 1100 °C. The arrows represent the scattered energies from the indicated elements if located at the surface.

Whereas only a portion of the underlying polysilicon is consumed in forming CoSi at 600 °C, the Co and polysilicon are almost completely reacted to form the disilicide by 800 °C. Even after 30 min at 1100 °C, the slope of the top, the total width of the Co peak, and the low-energy edge of the Co signal remain unchanged from that at 800 °C, showing reasonable stability of the silicide at 1100 °C. However, the leading Co edge appears to have receded from the surface by 600 Å, suffering a simultaneous decrease in slope. In addition, the Co peak height appears to have increased from that at 800 °C by 5%, possibly due to silicon loss, and a small oxygen surface peak is clearly detectable. The interaction between Co and (100) Si closely resembles that between Co and polysilicon.

In order to check if these high-temperature characteristics of the Co–Si interaction are retained when the underlying polysilicon is only partially transformed to $CoSi_2$, we sputtered a thinner Co layer (400 Å) onto polysilicon (3300 Å) and processed it through the same range of temperatures in forming gas (FG). The corresponding RBS spectra are shown in Figure 3.15. After 30 min at 900 °C, the leading edge of the Si signal appears at the surface and the front half of the polysilicon transform into the disilicide, substantiating the results of Figure 3.14. The disilicide grows as a uniform layer on polysilicon, as evidenced by the formation of plateaus in the Si backscattering signal. However, after 30 min at 1000 °C, the Co

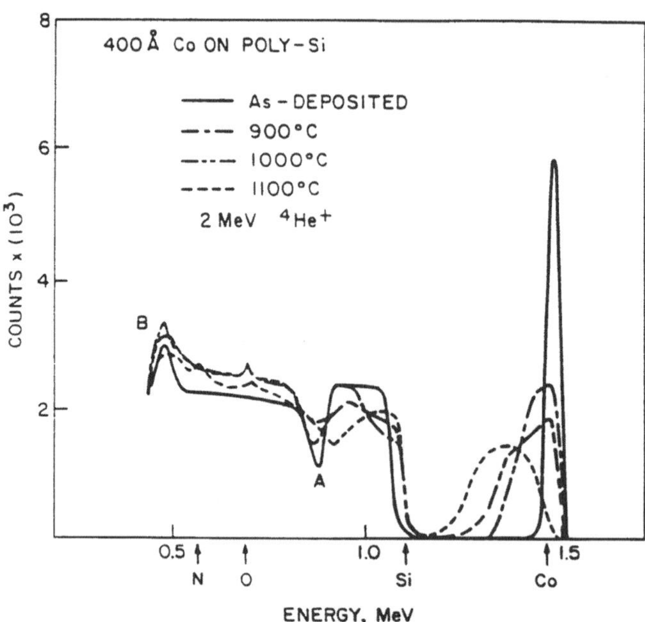

Figure 3.15 **Superimposed RBS spectra of 400 Å Co on polysilicon, as-deposited, and after 30 min forming gas anneals at 900, 1000, and 1100 °C. The dip A and the peak B correspond to Si and O, respectively, in the 1000-Å-thick SiO$_2$ layer under polysilicon (also true for Figure 3.14).**

surface concentration decreases and the metal is no longer localized at the surface but extends into the underlying polysilicon. This is complemented by an enrichment of Si at the surface and a continuously varying Si profile in the silicide–silicon composite. With increasing temperature (1100 °C), the Co peak becomes further diffuse and recedes from the surface, whereas the Si surface concentration progressively increases. In addition, surface oxygen and nitrogen can be identified at temperatures ≥900 °C. Thus, by 1100 °C the silicide appears to be covered by a 600–800-Å-thick layer of Co–Si–O–N. It should be noted that the "dip" in the Si spectrum (labeled A), arising from the underlying 1000-Å SiO$_2$ layer, progressively increases in height with increasing temperature in both Figures 3.14 and 3.15. The corresponding oxygen peak in Figure 3.15 (labeled B), however, appears to remain unaltered up to 1000 °C. These results demonstrate relative instability of these CoSi$_2$ films compared to thick CoSi$_2$ films, formed by reaction of metal with cobalt and heat-treated in a FG ambient.

The stability of cobalt disilicide is improved if the postannealing heat treatments are carried out in oxidizing ambients. The results are best described by RBS profiles of codeposited cobalt disilicide on oxide (in this case called wet oxide [WOX] for oxides grown in steam). Figure 3.16 clearly demonstrates the role of the annealing ambient. The disilicide formed after a 900 °C/30-min/FG sinter was heated in a

  SILICIDES Chapter 3

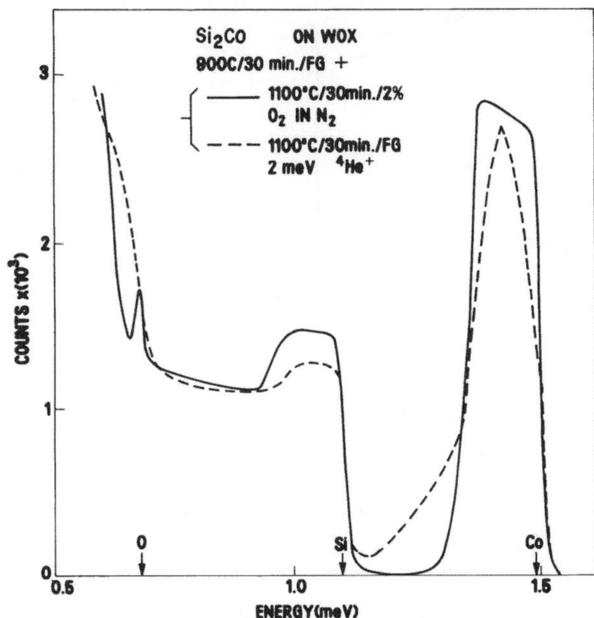

Figure 3.16 RBS spectra of cosputtered (Si/Co = 2.2) films on 5000 Å wet-oxide (WOX) after 1100 °C/30 min/2% oxygen in N$_2$ and 1100°C/30 min/ forming gas anneal, which preceded the 1100 °C anneals.

2% oxygen-in-nitrogen environment for 30-min intervals at 1000 and 1100 °C. Except for the detection of surface oxygen in the case of the oxidizing anneals, the 1000 °C spectra of thin-film Co as well as cosputtered "CoSi$_{2.2}$" on polysilicon, appear identical to the corresponding profiles in FG. Higher temperatures (1100 °C) merely increase the thickness of the surface oxide for cobalt-on-polysilicon composites. For cosputtered films, on the other hand, the apparent interdif-fusion of different atomic species after a 1100 °C treatment can be prevented by annealing in oxygen. When the RBS spectra of cosputtered "CoSi$_{2.2}$" on thick oxide after the 900 °C/FG anneal and the 1100 °C/2% O$_2$ anneal are superposed, they appear similar except for the addition of a 300-Å surface oxygen peak in the latter. In contrast, an equivalent 1100 °C sinter in FG results in a highly nonuniform Co distribution and loss of Si from the silicide.

The role of the annealing ambient in the formation of titanium silicide by reaction of Ti with Si is demonstrated in Figure 3.17. Iyer et al.[42] have shown that the Ti–Si reaction is slower when annealed in He compared to the anneals in N$_2$ or Argon. They interpret this difference to be due to the ability of helium, a small atomic species, to stuff the Ti grain boundaries very quickly, thus slowing down the Si diffusion, through grain boundaries in Ti film.

The RBS technique can employ other ion beams for improving resolution. For example, Figure 3.18 shows titanium silicide formation using rapid thermal annealing.[43]

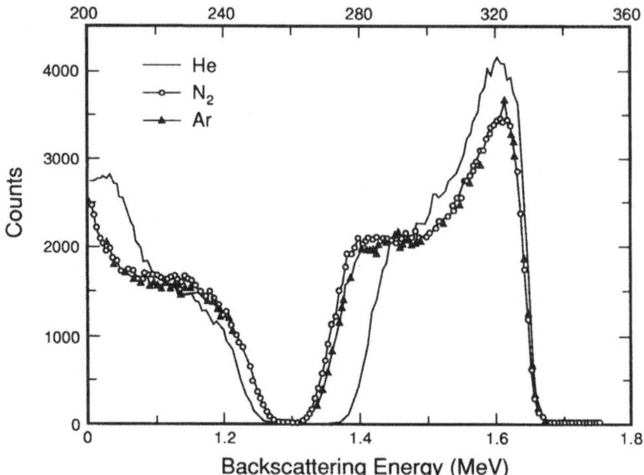

Figure 3.17   RBS spectra for 200 nm Ti on Si annealed at 600 °C for 30 min in He and N₂. (From Reference 42.)

The RBS spectra were obtained using a 16 MeV ^{16}O beam. Spectra from different processing times are superimposed. The 3 s anneal formed 360 Å of $TiSi_2$. The silicide formation was complete after 8 s anneal, approximately 1420 Å $TiSi_2$ is formed. Note in this study no titanium rich compounds were detected. These studies demonstrate the direct formation of $TiSi_2$ by rapid thermal annealing.

X-Ray Diffraction Measurements

Both X-ray and electron diffraction provide information about the microstructure and crystallinity of films, the information not obtained by the resistance or RBS

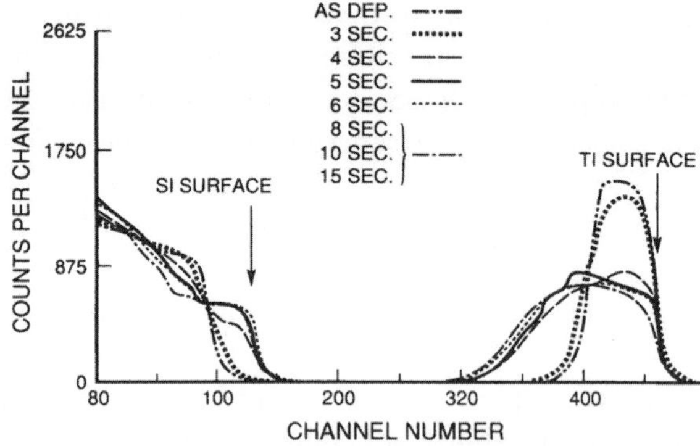

Figure 3.18   RBS spectra superimposed for different processing times. A 16-MeV ^{16}O beam was used to increase the mass resolution. The samples were positioned normal to the beam. The backscattered O particles were detected at an angle of 110° with respect to the incident beam by a Si surface-barrier detector. (From Reference 43.)

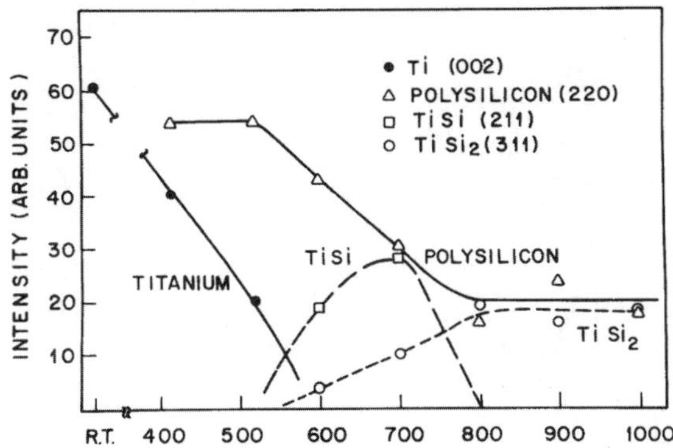

Figure 3.19 The XRD (integrated) intensities of Ti, polysilicon, TiSi, and TiSi$_2$ as a function of the vacuum sintering temperature.

measurements. Electron diffraction requires difficult sample preparations, making XRD a more practical day-to-day tool. However, following the kinetics of a phase formation is more demanding, especially where more than one phase forms simultaneously. Figure 3.19 shows the integrated intensities of various diffracted lines representing Ti, polysilicon, TiSi, and TiSi$_2$ as a function of the alloying temperature. As is seen, Ti is rapidly consumed with increasing temperature and no Ti is detected at 600 °C or above. Polysilicon is consumed up to 800 °C. TiSi and TiSi$_2$ are first detected after a 600 °C anneal. The amount of TiSi first increases with increasing temperature and then decreases and is not detected at 800 °C. The amount of TiSi$_2$ increases up to 800 °C and then, as inferred from the intensity of the XRD peaks, becomes independent of anneal temperature. At about 700 °C all of the Ti is consumed to form TiSi. Between 600 and 800 °C TiSi reacts with the underlying polysilicon and forms TiSi$_2$, thus accounting for the decrease in the intensity of the polysilicon X-ray diffraction peak and increase in the intensity of the TiSi$_2$ peaks. Finally, at 800 °C or higher temperatures, complete conversion to TiSi$_2$ has occurred, leading to no further effect on polysilicon or TiSi$_2$ diffraction peak intensities. Sheet resistance measurements have also been used to monitor the kinetics of formation of silicide in codeposited films of Ti–Si. The activation energy of TiSi$_2$ formation in the codeposited silicide is found to be the same as obtained from interaction between Ti and Si, indicating a similar thermally activated process in two types of sample preparations.

Figure 3.20 shows that the reaction of Co with polysilicon eventually leading to CoSi$_2$ formation is considerably retarded when the polysilicon is heavily doped with phosphorus. CoSi diffracton peaks dominate the cobalt-on-phosphorus-doped polysilicon. On the other hand, CoSi$_2$ peaks are the predominant ones in cobalt-on-undoped-polysilicon annealed under identical conditions. In this case, XRD and R_s measurement went hand-in-hand extremely well.

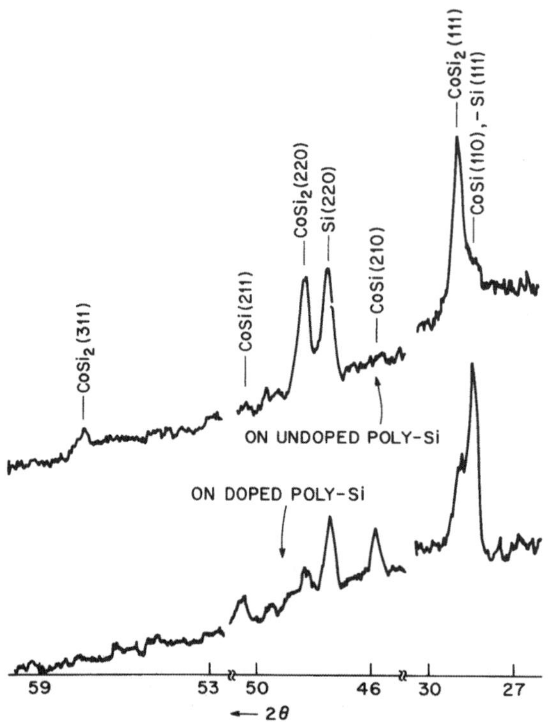

Figure 3.20 The X-ray diffractometer traces obtained from cobalt-on-undoped and -doped polysilicon substrates. Samples were annealed for 30 min in FG at 600 °C. Only significant parts of the total traces are shown.

Figure 3.21 shows a plot of the R_s and the X-ray intensity of the cosputtered W–Si (with a Si/W atomic ratio of 2.2) film on SiO_2 as a function of the annealing temperature. Initially, the R_s measurements appeared to be anomalous, but the XRD studies resolved the mystery by indicating the presence of a new and metastable hexagonal phase of WSi_2.[44]

Ellipsometric Measurements

Because of the resolution limits of the conventional XRD, RBS, SIMS, or AES coupled with ion milling, the study of the reaction between thin (<300 Å) metal films and silicon would be difficult using these techniques. The resistance measuring technique is also inadequate because of the thickness dependence of the resistivity in extremely thin films.[45] Ellipsometry has, therefore, recently been used to investigate the kinetics of the formation of $CoSi_2$ using 100- to 300-Å cobalt films on silicon and a rapid thermal annealer (RTA).[31] By use of such thin films, one can eliminate the effect of the larger extinction coefficient (associated with metals and metallic alloys), since the light is able to reach the substrate and reflect back. Figure

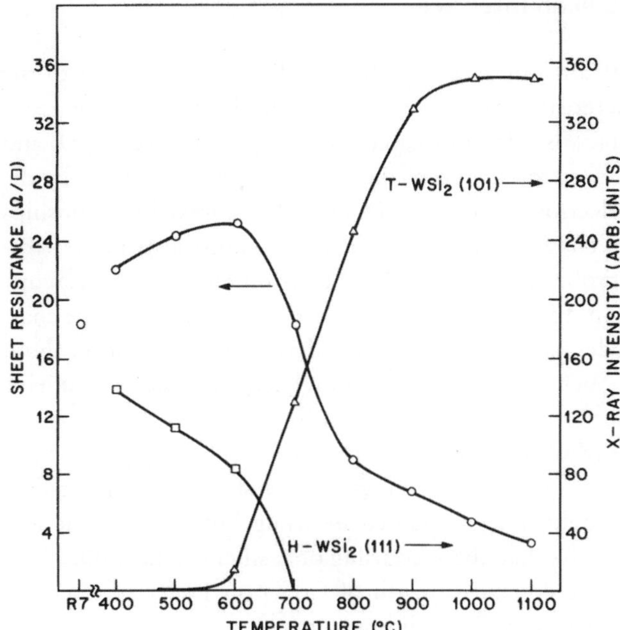

Figure 3.21   Sheet resistance and the integrated XRD intensity (of both hex. WSi_2 and tet. WSi_2) as a function of sintering temperature (30 min in 15% H_2/85% N_2). The film was nearly 2500-Å-thick co-sputtered tungsten–silicon mixture (Si/W atomic ratio of 2.2).

3.22 shows the results of the study of the formation of $CoSi_2$ from reaction between CoSi and Si (CoSi was first formed using ~200-Å cobalt films).

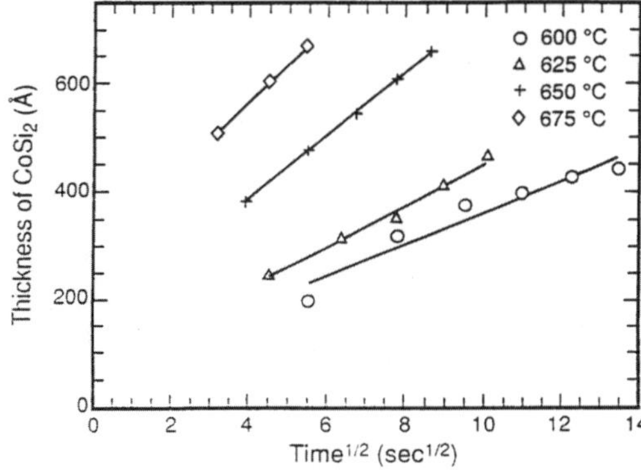

Figure 3.22   The thickness of $CoSi_2$, formed by interaction of CoSi with Si and determined by ellipsometric method, as a function of the square root of treatment time in an RTA at different temperatures.

3.3 The Silicide–Silicon Interface

The nature of the silicide–silicon interface, which contributes significantly to the Schottky barrier diode characteristics, contact resistance, and leakages in the active device under the silicide contact, is difficult to ascertain using resistance, XRD, and even the RBS measurements. A pictorial view of the interface can be obtained by the cross-sectional transmission electron microscopy (XTEM)[46]. Epitaxial relationships can be examined by the use of channeling RBS.[35] The X-ray photoelectron spectroscopy (XPS),[47] Raman microprobe analysis,[48] ballistic electron emission microscopy (BEEM),[49] low-energy electron diffraction (LEED),[50] and the current–voltage characteristics of the Schottky diodes have also been used to derive the information about the silicide–silicon interface. Such information is then related with the bulk silicide properties.

Figure 3.23 shows the XTEM of the $CoSi_2$ on polysilicon.[46] The bright and dark-field images of the $CoSi_2$ on polysilicon are shown: 3.23*a* shows the bright field image, 3.23*b* shows the dark-field image obtained by using both the silicide and silicon reflections, and 3.23*c* shows that obtained using only silicide reflection. These micrographs clearly establish the epitaxial growth of $CoSi_2$ on preferred grains of polysilicon and supports the lower stress in such structures.[16]

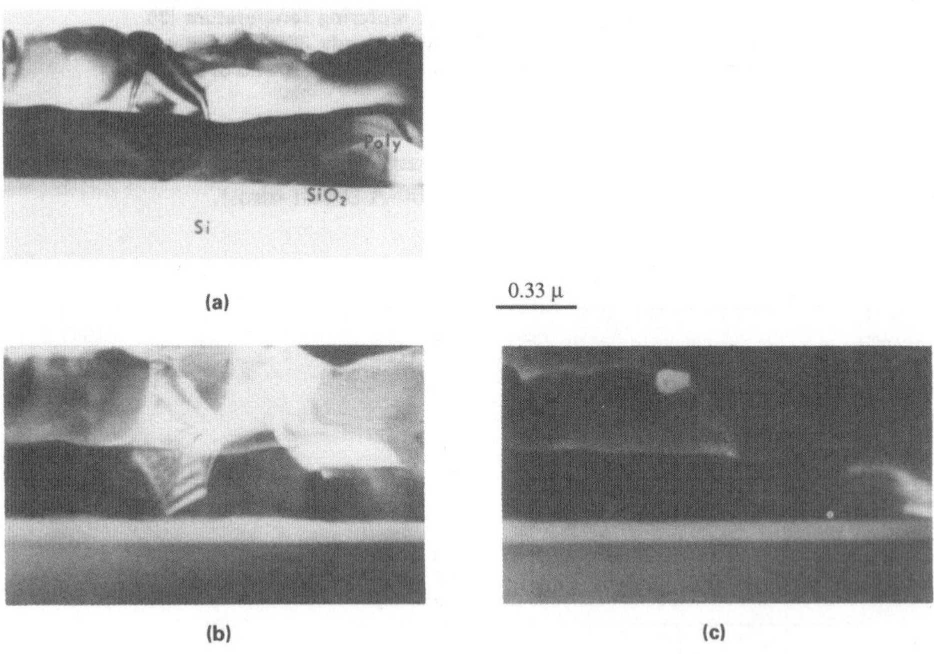

Figure 3.23 XTEM micrographs of cobalt-on-polysilicon sample annealed to form the $CoSi_2$, showing epitaxial growth of $CoSi_2$ on preferred polysilicon grains: (*a*) bright-field image, (*b*) dark-field image using both silicon and silicide reflections, and (*c*) dark-field image using silicide reflection only.

When the incident ion beam direction is aligned to match the crystal direction of so-called open channels between atoms, the ions can travel very large distances before being scattered significantly. Called channeling, this can be used as a tool to reveal the crystalline structure.[35] Analyzing the RBS spectra from the samples that are rotated and measuring the ion-channeling yield in random and channeling directions provide very convenient means of identifying epitaxial growth. If the film consists of randomly oriented crystallites, the spectra will nearly all be the same. On the other hand, if crystalline growth exists, at certain angles the crystalline film will align parallel to a major crystal axis and to the incident ion beam. The yield of scattered ions will then be considerably reduced. The measure of crystallinity has been defined in terms of a quantity χ_{min}, called channeling. The channeling χ_{min} is defined as the ratio of the minimum channeling yield for the aligned specimen to the total random yield for the nonaligned case.

Figure 3.24, for example, shows the RBS spectrum of a randomly oriented 750-Å $NiSi_2$ film formed by deposition of 190-Å Ni on a 55-Å-thick $NiSi_2$ template layer heated at 650 °C.[51] Nearly perfect epitaxy with a χ_{min} of 3% along the (110) axis is seen. Both $NiSi_2$ and $CoSi_2$ form excellent epitaxy on (111) silicon, and channeling RBS provides a very quick and nondestructive means of evaluating the epitaxial perfection. For MBE, $CoSi_2$, χ_{min} was found to be 0.02, which is among the

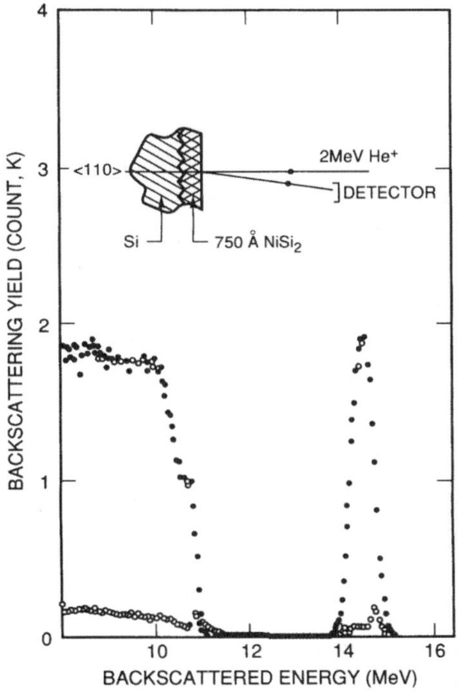

Figure 3.24 Channeling (*open circles*) and random (*closed circles*) RBS spectra of a 750-Å-thick $NiSi_2$ layer on Si (110). (From Reference 51.)

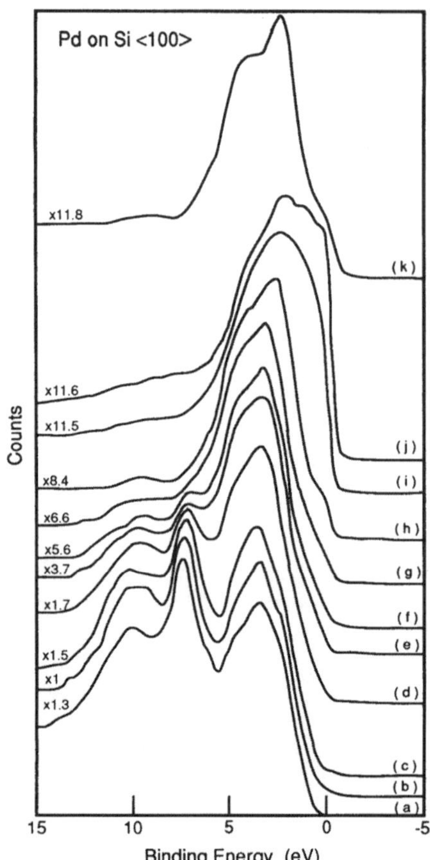

Figure 3.25 Plot of the evolution of the XPS valence band spectra as increasing amounts of Pd are deposited on Si <100> at room temperature. In (*k*), the sample has been heated to 473 K to yield Pd_2Si. (From Reference 47.)

best reported for crystalline materials.[52] Such epitaxial silicides and heteroepitaxy of silicon on epitaxial silicide will, perhaps, play a major role in the future in fabricating three-dimensional and fast devices.

A considerable understanding of the reactivity, chemical bonding, and electronic characteristics of the metal–silicon interface can be obtained by use of high-resolution XPS. For example, a plot of the evolution of the XPS valence bond spectra as more and more Pd is deposited on Si is shown in Figure 3.25 together with that of a Pd-on-Si sample heated to 473 K to form Pd_2Si. Grunthaner et al.,[47] who obtained these results, point out that a shift in the main peak is observed in (*h*), (*i*), and (*j*), representing the presence of pure Pd metal. However, on silicide formation a further shift occurs associated with charge transfer from Pd to Si, indicating a silicon-rich environment. XPS studies have indicated that the metal–silicon

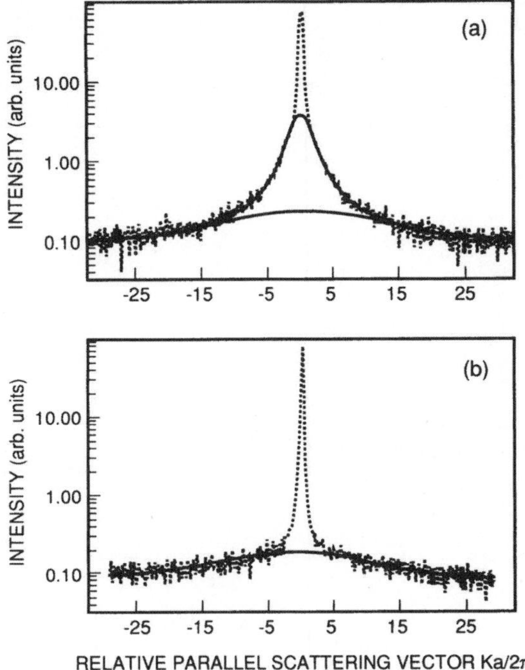

Figure 3.26 **The spot profile after evaporation of four monolayers Ni on Si (111) and annealing at 630 K: (a) out-of phase condition—the broadening is composed of two Lorentzians (*solid lines*) and (b) in-phase condition—only the broad Lorentzian (*solid line*) remains. (From Reference 54.)**

interface (even in the as-deposited condition) is graded in composition, being silicon-rich (relative to Pd_2Si) next to crystalline silicon.[47]

Very recently, BEEM spectroscopy[49] has been used to investigate carrier transport in the epitaxial $CoSi_2/Si$ system.[50] BEEM is a new technique for measuring ballistic electron and hole transport across The metal–semiconductor Schottky barrier interfaces with high-energy and spatial resolution.[53]

The initial stage of the growth of the silicide at the silicide–silicon interface is not very well understood. When a few monolayers of metal is deposited on a clean silicon surface, the silicide growth will most likely occur by (1) island growth on the surface leading to surface asperities, (2) silicide growth in localized areas below the silicon surface, and (3) uniform growth on or under the silicon surface. Generally, TEM and scanning tunneling microscopy (STM) have been employed to examine such surfaces. Very recently, high-resolution LEED has been used to investigate the asperities and the inhomogenerties on the surface of silicon during silicide formation.[54] Such inhomogenieties at the surface produce an additional broadening of the diffraction spots as shown in Figure 3.26, which shows the LEED profiles after the annealing of four monolayers of Ni on (111) Si at 630 K to form $NiSi_2$.[54] Wollschlager et al.[54]

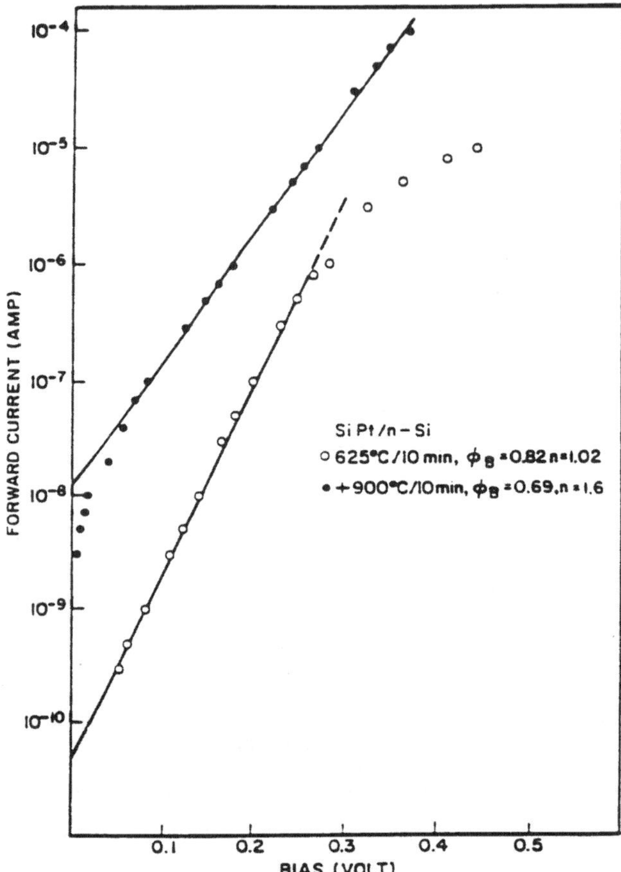

Figure 3.27   Forward I–V characteristics of SiPt/n-Si Schottky diode after 625 and 625 + 900 °C anneals.

conclude, "The wide broadening is undoubtedly produced by inhomogeneities of the surface."

Simple current–voltage (I–V) characteristics of the silicide–silicon Schottky diodes are very useful in elucidating the stability of the silicide–silicon interface and diffusion across the interface. Figure 3.27 demonstrates the instability of the PtSi/n-Si interface when annealed at temperatures higher than 700 °C (in this case 900 °C). The higher temperature anneal led to nonideal and leaky diodes, indicating a deeper diffusion of Pt into Si. Figures 3.28 and 3.29 show the stability of the Al/CoSi$_2$/n-Si structure up to temperatures of ~450 °C and penetration of Al through CoSi$_2$ to the silicide–silicon interface leading to an increased Schottky barrier height and ideality factor.[55] Note in this case the diodes carry less current after a 500 °C anneal because the Al/n-Si Schottky diode has a higher barrier height. These results electrically establish the stability of the CoSi$_2$/Si structure

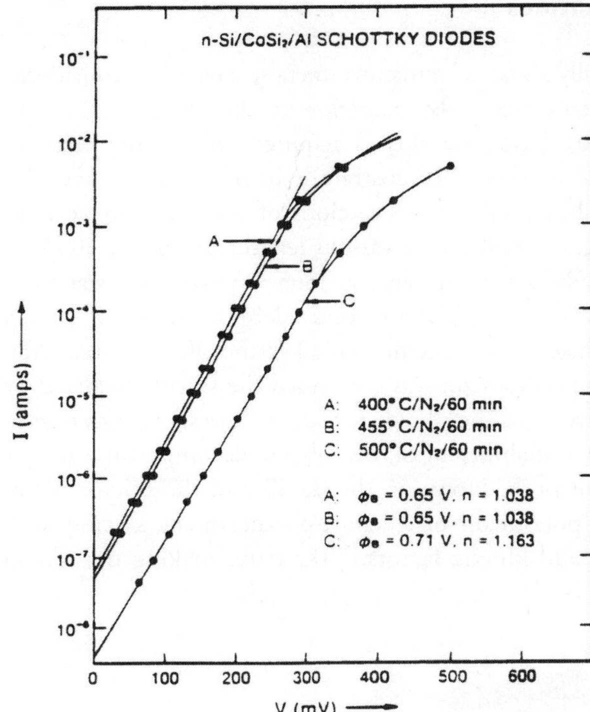

Figure 3.28   Forward I–V characteristics of Al/CoSi$_2$/n-Si Schottky diodes after various anneals.

against aluminum penetration and are supported by RBS, sheet resistance, and Auger analyses.

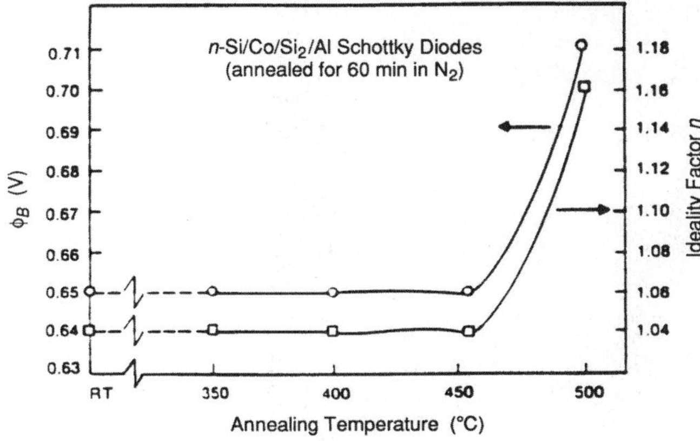

Figure 3.29   Schottky barrier height and ideality factor versus annealing temperature for Al/CoSi$_2$/n-Si devices.

3.4 Oxidation of Silicides

Although bulk silicides, especially those of refractory metals, have been considered corrosion- and oxidation-resistant materials, oxidation studies of thin-film silicides have clearly shown that oxidation stability is acquired only in the presence of excess silicon in the silicide, or when the substrate is in intimate contact with the silicide layer. Figure 3.30 shows TEM cross sections of the $TaSi_2$ on polysilicon (on thermally oxidized silicon) oxidized for various lengths of time at 1050 °C in steam.[56] It is clear from the cross sections that, as long as polysilicon was present (i.e., for shorter oxidation times), oxidation occurred by transport of silicon through the silicide to the surface of the silicide. This left the silicide intact (top two cross sections). Once all the polysilicon was consumed the silicide oxidized by decomposition, leading to a decrease in the silicide thickness (lowest cross section in Figure 3.30). In absence of the availability of Si, the silicide decomposition occurs by the oxidation of metal, Si, or both. However, the stability of the silicide is not guaranteed by the presence of polysilicon or silicon underneath the silicide. It is also related to thermodynamic and kinetic factors,[15] the latter making the silicon

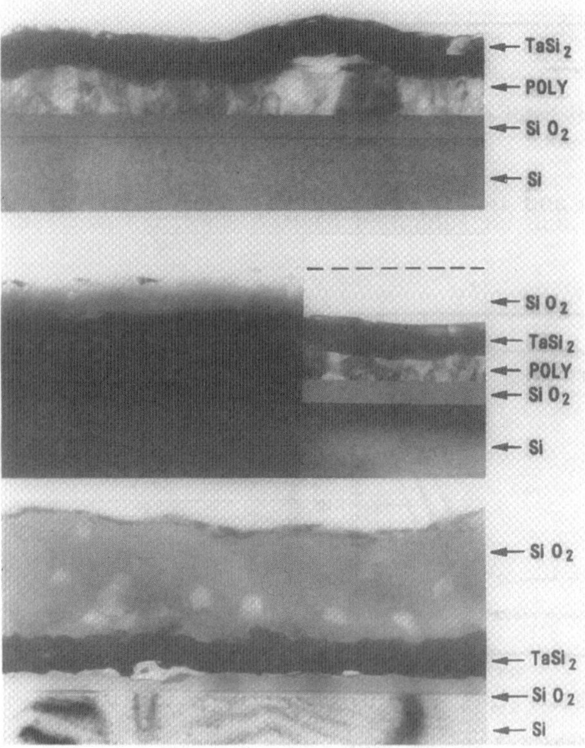

Figure 3.30 **TEM cross section of $TaSi_2$ (on polysilicon) oxidized for various lengths of time at 1050 °C in steam.**

available at the silicide–oxidizing species interface. For example, $HfSi_2$ is not stable in an oxidizing ambient even when silicon is available under the silicide. Figure 3.31 compares the Auger depth profiles of $HfSi_2$ on Si before and after oxidation.[57] It is clear that all of the $HfSi_2$ film is oxidized in 20 min at 700 °C in oxygen. One can explain the difference between $HfSi_2$ and $TaSi_2$ by noting that the free-energy of formation of Hf oxide is very large compared to that of SiO_2 and Ta oxide. Lowering of the free energy thus drives the Hf to oxidize readily.[58]

In spite of the above understanding, the oxidation mechanism in the silicide–silicon system is not clearly established. Several published papers report oxidation

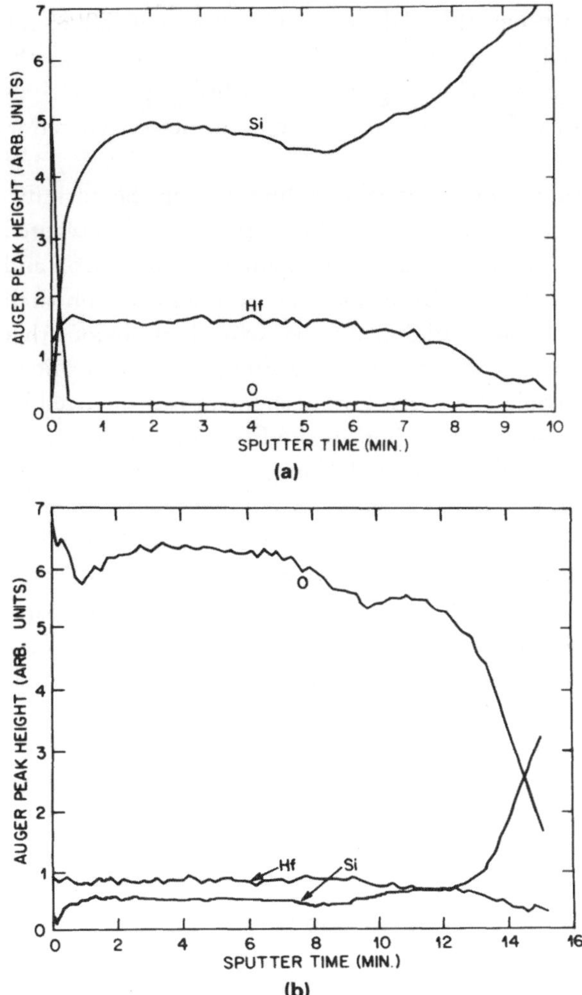

Figure 3.31 Auger depth profiles of a hafnium silicide film on silicon substrate formed (*a*) after a 900 °C anneal of Hf on Si in vacuum for 30 min and (*b*) after oxidation at 700 °C for 20 min in oxygen.

characteristics of silicides, and there is considerable variation in reported results. It has only recently been realized that a large number of parameters determine oxidation rates and the quality of the oxides produced. Chu et al.[59] studied the effect of silicide deposition parameters—such as stoichiometry, substrate bias during cosputtering, and the substrate rotational speed—on the quality of the thermally grown oxide after deposition and anneal. Gant et al.[60] report the effects of dopants and deposition parameters—such as substrate heating, substrate cleaning prior to loading in the sputtering chamber, and substrate cleaning in situ by back-sputter etching—on the properties of the oxide thermally grown on cosputtered $TaSi_2$. They used sheet resistance, X-ray microprobe analysis, XRD, XTEM, and dielectric strength (I–V) measurements in these studies. Liu et al.[61] also studied oxidation rates as a function of the dopants and excess silicon using AES and step height measurements in the $TaSi_2$. The oxidation rate decreased with increasing silicon content in the film.[61, 62] These results indicate that the excess silicon plugs the fast diffusion paths such as dislocations and grain boundaries and perhaps reduces the point defect generation in the silicide lattice.

Dopants in underlying polysilicon or in the silicide film also appear to influence the oxidation rate.[61] Boron implanted in polysilicon and arsenic implanted in $TaSi_2$ seem to enhance the oxidation rate. Arsenic implanted in polysilicon and PBr_3 doping of polysilicon retard the oxidation rate. These rates are compared with oxidation rates of undoped $TaSi_2$ on undoped polysilicon. It is obvious that the determination of a representative oxidation rate for a silicide film on a silicon or polysilicon substrate is not trivial. The oxidation mechanism is influenced by the silicon content and the dopant concentration in the film and the substrate, in addition to the grain size in the film. The oxidation process in silicides over silicon is similar to that of silicon. The oxide thickness is given by the famous Deal–Grove model, in which a linear oxidation rate fits at early stages of oxidation and a parabolic oxidation rate occurs at larger thickness. For silicides, the linear oxidation rate appears to depend on the type of silicide, whereas the parabolic rate seems to be independent of silicide type. A careful analysis of the data is necessary. Several attempts have been made to formulate a mechanism that explain all oxidation behaviors.[59, 61, 63]

Chu et al.[59] and Gant et al.[60] have clearly shown that routine oxidations to form oxides (as thick as 5000 Å) on silicides can be carried out readily. The oxide–silicide interfaces look excellent and voids were not found to be an intrinsic characteristic of the oxide growth.

3.5 Dopant Redistribution During Silicide Formation

In silicon IC applications, defined at the beginning of this chapter, the silicides are used on top of the silicon substrates or on the thin polysilicon films. In both cases the silicon is doped, generally to very high concentrations (10^{19}–10^{21} atoms/cm^3) with boron, phosphorus, or arsenic. During the silicide formation, which is induced by annealing an as-deposited silicon–metal mixture or by reacting a metal on silicon or polysilicon, dopants are found to redistribute between the silicide and the substrate, leading to a reduction in the dopant concentration in the substrate and to an incorporation of the dopant in the silicide.[63] The results reported in the literature show a variety of behaviors: (1) snowplowing occurs, in which the silicide formation pushes the dopant deeper into the silicon with little or no diffusion occurring in the silicide; (2) dopant redistribution occurs uniformly throughout the silicide; (3) dopant pileup occurs at the silicide–silicon interface and at the silicide surface; and (4) dopant diffusion occurs through the silicide and loss to the surrounding medium results because of evaporation or reactive losses. In addition, the presence of dopants and their concentration significantly influence the kinetics of the metal–silicon reaction. These effects must be considered seriously because the dopant redistribution will affect the electrical properties of the silicon or polysilicon layers. This, in turn, will affect the device performance. Thus, understanding of these dopant effects is essential.

RBS can be useful in analyzing the dopant profiles in the silicon–silicide system for all dopants except for boron and phosphorous. Most of the studies, however, involved arsenic as the dopant. Figure 3.32 shows the redistribution of arsenic, implanted into a CVD $WSi_{2.6}$/Si sample, as a result of a furnace and a rapid thermal

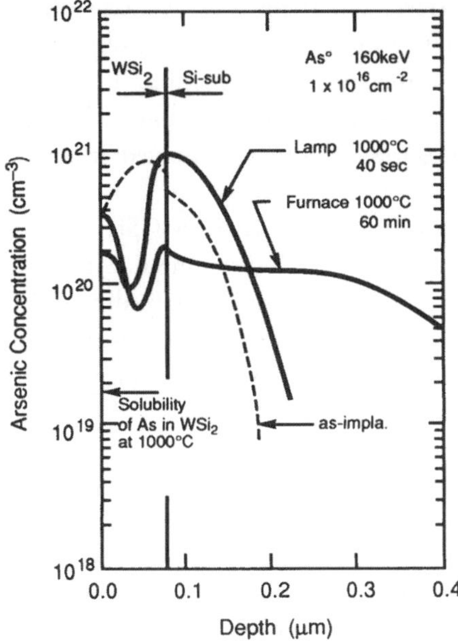

Figure 3.32   Impurity concentration profiles of arsenic implanted CVD $WSi_{2.6}$ (800 Å)/Si, where arsenic was implanted at 160 keV. Annealing was done at 1000 °C by furnace heating and RTA for 60 min and 40 s, respectively. Broken curve shows as-implanted profile. Solid solubility of arsenic in $WSi_{2.6}$ at 1000 °C is also shown. (From Reference 64.)

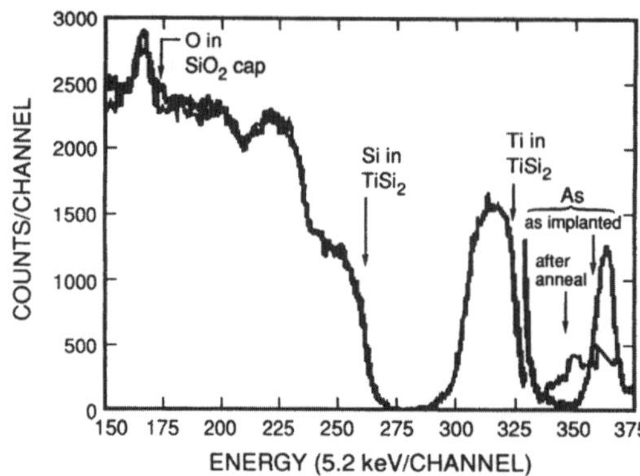

Figure 3.33 RBS spectra for a TiSi₂/polysilicon bilayer which had arsenic implanted into the TiSi and was then annealed at 850 °C. (From Reference 65.)

anneals.[64] Arsenic rapidly diffused out of the silicide into silicon, the diffusion being highly pronounced during furnace anneal compared to rapid thermal anneal. The results also indicate that the As concentration in the silicide at the silicide–silicon interface is higher than that in the bulk of the silicide. Similarly, the As concentration is higher in the silicide volume close to the surface. The latter is associated with the tendency of arsenic to evaporate into the annealing ambient, whereas the segregation at the silicide–silicon interface is related to the segregation phenomenon. A comparison of RTA and furnace-annealed samples suggests that the furnace-annealed sample is closer to thermal equilibrium and any segregation coefficient derived from this sample will be more realistic.

Figure 3.33 shows the arsenic redistribution of implanted As into TiSi₂ on polysilicon.[65] In contrast to the case of WSi₂, in this case As redistributed within the silicide without any significant out-diffusion into the underlying polysilicon film. These authors also found that arsenic initially implanted into polysilicon migrated to the covering titanium silicide layer, confirming the extremely high affinity of TiSi₂ for arsenic, as demonstrated in Figure 3.33.

For the boron and phosphorus redistribution studies, Auger or secondary ion mass spectrometry (SIMS) coupled with ion-milling are used. Figure 3.34 shows that after the silicide formation anneal of titanium on silicon at 700 °C for 30 min, the originally placed boron in silicon (using BF_2^+ implantation) has redistributed leading to (1) a peak in the boron concentration at the silicide–silicon interface and (2) a peak at the silicide surface. The latter indicated a tendency of loss from the surface to the ambient.[66] In this study it was also shown that the peak concentration at the silicide–silicon interface decreased with increasing anneal temperature. The effect of temperature is perhaps associated with the continued consumption of silicon and accompaning transformation from titanium-rich silicide at 650 °C to near-stoichiometric silicide at higher temperatures.

SILICIDES   Chapter 3

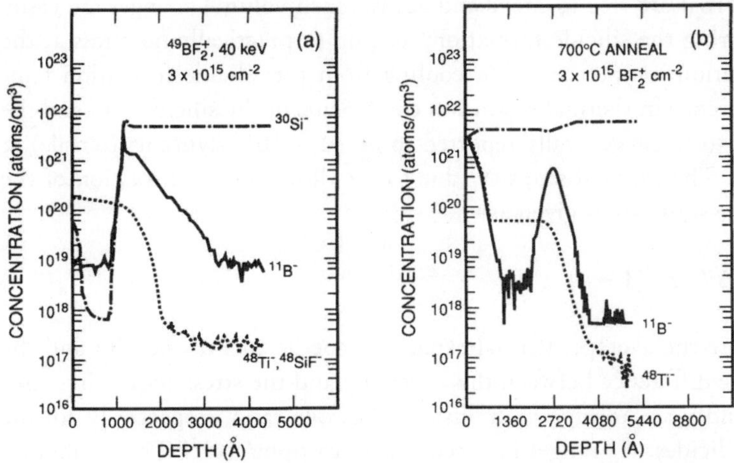

Figure 3.34 SIMS depth profiles for a Ti/Si sample implanted with 3×10^{15} BF$_2$ cm^{-2} ions (*a*) before and (*b*) after annealing at 700 °C for 30 min in H$_2$. (From Reference 66.)

Dopant diffusion in silicides is also of significant concern during processing of the codeposited silicides. Dopants in the silicide or the underlying layer or the substrate redistribute between the silicide and the substrate. Using Auger analysis coupled with ion-milling, Tsai et al.[67] have shown that the redistribution of the phosphorus between the phosphorus-doped polysilicon and WSi$_2$ layers led to significant loss of phosphorus from the polysilicon, in turn leading to a threshold voltage change in the metal/oxide/semiconductor capacitor. Pan et al.[68] found boron redistribution to be quite different from redistributions of phosphorus and arsenic in WSi$_2$. Boron piled up in WSi$_2$, whereas phosphorus and arsenic seem to redistribute differently in different films, with a dopant solubility in WSi$_2$ of about 6×10^{19} P atoms cm^{-3} and 1.6×10^{19} As atoms cm^{-3} at 1000 °C. Vaidya et al.[69] have reported results of boron, phosphorus, and arsenic redistribution in TaSi$_2$/polysilicon structures. Once again, boron piled up in TaSi$_2$, and the behavior of phosphorus and arsenic was similar to that in WSi$_2$. Pelleg and Murarka,[70] using neutron activation analysis (NAA) and SIMS, determined the diffusion of phosphorus in TaSi$_2$ (cosputtered and sintered). They also found very high diffusion rates of approximately 2.5×10^{-12} cm^{2}s^{-1} at 1000 °C; a very similar value, of 3.3×10^{-12} cm^{2}s^{-1} at 1000 °C, has been reported[68] for dopant diffusion in WSi$_2$. Such high diffusivities, several orders of magnitude larger than in silicon, indicate contributions not only from the grain boundaries, but also from the dislocations in the silicide grains.

3.6 Stress in Silicides

Silicides form by metal–silicon reaction which causes a loss of volume since the product volume is less than the total reactant volume. However, since the reaction occurs on surfaces that are free to move and adjust, such volume changes are easily accommodated during the silicide formation, leading to practically no stress at the high silicide formation temperature. On cooling from the silicide formation temperatures, the difference in thermal expansion coefficients of the silicide film and the substrate will lead to stress generally reported to be ~1 × 10^{10} dyn/cm^2 (tensile) at room temperature. The thermal stress σ_T due to the differential contraction of the silicide film and the substrate is given as

$$\sigma_T = (\alpha_F - \alpha_s)\Delta T \cdot E/(1 - v) \tag{3.4}$$

where α_F and α_s are the average thermal expansion coefficients of the film and the substrate, ΔT is the difference between the annealing and the stress-measuring temperatures, and E and v are Young's modulus and Poisson's ratio, respectively, for the silicide film. For silicides, α_F ranges between 6 and 15 ppm/ °C.[16] This is about a factor of 2 to 5 higher than the α of silicon, giving a σ_T in the range of 3–12 × 10^9 dyn/cm^2 for a silicide anneal temperature of 1000 °C and E = 1 × 10^{12} dyn/cm^2. In spite of such high stresses, the films are very stable as long as the silicide–substrate interface is free of impurities that reduce adhesion.

For a codeposited film, silicide formation at lower temperatures leads to a high tensile stress at that temperature. On continued annealing to higher temperatures, the stress decreases due to grain growth and thermal expansion. For cosputtered TaSi$_2$ films on doped and undoped polysilicon substrates, this type of stress versus temperature curve is shown in Figure 3.35.[71] Also shown is the stress on cooling. The stress rises to a value in the range given above. Subsequent thermal cycling does not change the stress–temperature relationship.

It is important to point out that codeposited silicides on GaAs are found to have much lower stress at room temperature.[72] GaAs's thermal expansion coefficient is

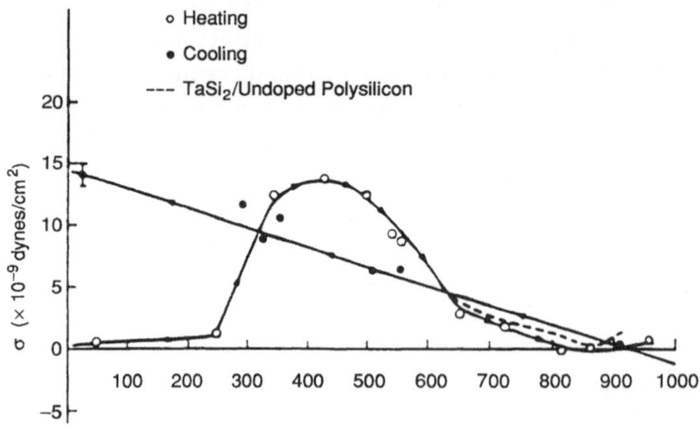

Figure 3.35   **Stress versus temperature curve for a sample with TaSi$_2$ deposited on large-grained (PBr$_3$-doped) polysilicon.**

very similar to that of silicides, and σ_T approaches the near zero value. In the absence of thermal stresses that dominate silicides on silicon or polysilicon substrates, the intrinsic stress due to lattice mismatch, defects, gases trapped in the film, etc., could be important. The A or B type of epitaxy in the case of $CoSi_2$ and $NiSi_2$ could result in a different state of stress that may be responsible for the variance in the barrier heights. σ_T is directly related to ΔT. With the continued shrinking of device dimensions and junction depths, the processing temperatures are decreasing. Thus, it is possible to conceive of temperatures at which the silicide stresses would be half of the presently reported numbers, especially for the near-noble metal silicides that can be formed at temperatures of 600 °C or lower.

Stresses in the films are most commonly measured by measuring the changes in the radius of curvature (R) of the substrate induced by the deposition of the film. Stress σ is calculated using the Stony equation

$$\sigma = \frac{E}{1-\nu} \frac{h^2}{6Rt} \tag{3.5}$$

where E, ν, and h are the Young's modulus, Poisson's ratio, and thickness of the substrate, respectively, and t is thickness of the film. It may be noted that Equation 3.5 is valid for $t << h$. Most commonly used methods for measuring thin-film stress make use of the cantilever beam technique that requires the measurements of the substrate radius before (R_1) and after film deposition (R_2). The effective radius (defined as $R = (R_1 - R_2)/(R_1 R_2)$ is substituted in the Stony equation to calculate the thin-film stress. The radii are, generally, measured by either of two methods: (1) the split-beam technique or (2) the scanning-beam technique.

In the split-beam technique, a laser beam is split into parallel beams at a distance L apart. The two beams are directed toward the substrate surface and are reflected back to two position-sensitive detectors. The reflected angles, $\theta_1 + \theta_2$ are determined by the position of the reflected beam on the detectors. The average radius is simply the beam-to-beam distance divided by the sum of both reflected angles, $R = L/(\theta_1 + \theta_2)$ The radius and, hence, the film stress can be measured repeatedly as a function of time as well as temperature.

The scanning-beam technique employs a movable assembly solid-state laser, mirror, and position-sensitive detector. In this case, the reflected angle, θ, is measured as a function of the distance x traveled along the substrate. The average radius is the derivative of the angle with respect to the distance: $R = d\theta/dx$. Using this method, one can calculate the substrate deflection since it is merely the integral of θdx.

In the absence of the equipment of the type described above, experimentally it is easy to use two parallel laser beams aimed at the wafer. The application of some simple geometry can be used to find R by measuring the distance between the two reflected beams.

One of the most accurate methods for determining stress is to determine changes in the lattice parameter and line broading using X-ray or electron diffraction techniques. The stress σ is given by

$$\sigma = \frac{E}{2\nu} \frac{a_0 - a}{a_0} \tag{3.6}$$

or

$$\sigma = \frac{E}{1 - \nu} \frac{a - a_0}{a_0} \tag{3.7}$$

In these equations, E is Young's modulus, ν is Poisson's ratio of the film, and a_0 and a are the lattice constants of the unstrained and the strained material, respectively. Equation 3.6 refers to the value of a perpendicular to the film plane, and Equation 3.7 refers to the case where a is the lattice constant in the plane of the film. The technique, although accurate, is time-consuming and requires a good understanding of the diffraction process and the equipment. Recently, specialized diffractometers, such as a generalized focusing diffractometer, have been designed to determine the strain and hence the stress in the films.

3.7 Stability of Silicides

In order to provide an application on a device/circuit, one must carefully evaluate the effect of the device processing on the properties of the silicides. Mochizuki et al.[12] have claimed a good degree of process stability for $MoSi_2$. Crowder and Zirinsky[13] have similarly reported good process compatibility of WSi_2 on polysilicon. The sheet resistance and the stress in the $TiSi_2$, $TaSi_2$, and $CoSi_2$[77] films have been measured as a function of the device processing steps. The stability of various interfaces in the silicide/silicon/oxide/substrate-structure by use of the RBS technique before and after complete processing have been compared. Figure 3.36 shows a plot of the sheet resistance and stress of sintered metal on polysilicon silicide as a function of the processing steps. There is no significant change in the sheet resistance of the silicides at the end of complete processing. For the silicides the stress was lowered somewhat due to the ion-implantation damage which caused a slight increase in the sheet resistance (Figure 3.36). The stress in $TaSi_2$ increased following the arsenic implant anneal and reoxidation steps. After the final step there was a decrease to a final value lower than that for the as-sintered film. Similar results clearly demonstrate the stability of other refractory silicides.

Use of silicides at the gate and interconnection level and in contacts requires stability with the top-level Al metallization. Practically all silicide films react with Al at about 450 °C.[16] Thus, when used as contact metallization with Al as the top metal, silicides require a diffusion barrier. The barrier must, however, be metallic with low resistivity. Nicolet[73] has discussed diffusion barriers and their importance in forming reliable contacts and metallization schemes. Citing the high chemical and thermodynamic stability of transition metal nitrides, carbides, or borides, he recommends the use of these compounds as the diffusion barriers. The compound

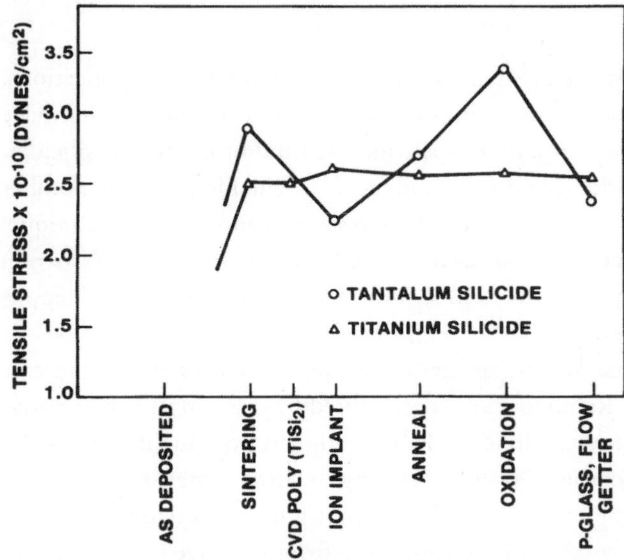

Figure 3.36   Stress (measured at room temperature) as a function of various process treatments given to TiSi$_2$ and TaSi$_2$ films.

films can easily be formed by sputtering from a compound target or by the reactive sputtering of the metal target in the appropriate gas (e.g., N_2 for nitride). The considerable amount of work that has been done in using nitrides and carbides as a diffusion barrier[73] between the silicide and aluminum indicates they seem to hold at least up to 450 °C.

The stability of the silicide during processing and during device operation is of utmost concern. In the last few years it has been recognized that stability is clearly related to the tight control required on the silicide deposition parameters, as discussed above. Beyond these concerns, each silicide has its limitation as originally discussed by this author.[15] Impurities, especially oxygen and carbon, play varying roles in different silicides. Silicides of titanium and tantalum (also those of Nb, V, Zr, and Hf) tolerate small amounts of oxygen without noticeable effect on mechanical stability. Other silicides behave differently. Silicides of tungsten and molybenum require special precautions, especially during oxidation. This is due to the inability of these metals to reduce the native interfacial oxide layer on silicon or polysilicon and to the reported occurrence of "pest reaction."[16]

Silicides of cobalt and platinum are only stable at temperatures less than 900 and 800 °C, respectively. Some results[74] indicate that the TiSi$_2$/polysilicon structure, in the presence of a thin oxide barrier between the polysilicon and TiSi$_2$, is not stable during 950 °C oxidation, and leads to penetration of titanium into polysilicon at the weak points in the interfacial oxide.

3.8 Summary

Studied for nearly a century, metal silicides have found a variety of applications. In this chapter we discuss the use of various characterization techniques that have been used in studying (1) the formation of silicides, (2) the properties of silicides, (3) the processing of the silicides, and (4) stability of the silicides. A great number of analytical techniques such as RBS, SIMS, AES, resistance monitoring technique, XRD, TEM, Schottky barrier height measurements, ellipsometry, and stress-strain measurements are used to elucidate the desired information. Examples of such applications are given and discussed.

Recent advances in and manufacturability of silicided structures clearly indicate that silicides are here to stay. Refractory silicides, limited by the ~50 $\mu\Omega\cdot$cm resistivity, will give way to low-resistivity silicides when the highest postsilicide processing temperatures are lowered to the 800–900 °C range. Three-dimensional devices with buried metal can only be made with silicides that grow epitaxially on silicon. Silicides for contact currently face competition only from selective CVD of tungsten which, when proven manufacturable, will lead to a simpler contact metallization scheme. Several new applications of silicides have been suggested in recent years. These include their use as: (1) a metallization for MESFET technology on GaAs where high-temperature and Schottky barrier height stability are necessary; (2) a base in metal-base bipolar transistors where formation of a thin epitaxial silicide on silicon followed by epitaxy of silicon on silicide is necessary; and (3) a ring contact to specially made bipolar transistors where selective area silicide formation is required. All such developments will follow research efforts which in turn will require the use of various analytical techniques discussed here. Newer developments may, however, need other more sensitive techniques that may or may not exist at this writing.

References

1 H. Moissan. *The Electric Furnace*. (English translation by A. de Mouilpied) Edward Arnold, London, 1904.

2 B. Aronsson. *Ark. Kemi.* **16**, 379, 1960.

3 R. Keiffer and F. Benesovsky. *Hartstoffe*. Springer-Verlag, Wien, 1963, p. 455.

4 R. Wehrmann. In *High-Temperature Materials and Technology*. (I. E. Campbell and E. M. Sherwood, Eds.) Wiley, New York, 1967, p. 399.

5 B. N. Padnos. "Chemical/Metallurgical Properties of Silicon." *Integrated Silicon Device Technology*. Vol. 10, Research Triangle Inst., NC, 1965, CA **66**, 80289.

6 K. E. Sundstrom, S. Peterson, and P. T. Tove. *Phys. Stat. Sol. (a)*. **20**, 653, 1973.

7   M. P. Lepselter and J. M. Andrews. In *Ohmic Contacts to Semiconductors.* (B. Schwartz, Ed.) The Electrochemical Society, Princeton, NJ, 1969, p. 159. Also, see U.S. patents 3,274,670, 27 Sept. 1966; 3,287,612, 22 Nov. 1966; and others by M. P. Lepselter.

8   D. Kahng and M. P. Lepselter. *Bell System Techn. Journal.* **44**, 1525, 1965.

9   J. C. Phillips. In *Thin Film Phenomena—Interface and Interactions.* (J. E. E. Baglin and J. M. Poate, Eds.) The Electrochemical Society, Princeton, 1978, p. 3.

10   K. N. Tu and J. W. Mayer. In *Thin Films—Interdiffusion and Reactions.* (J. M. Poate, K. N. Tu, and J. W. Mayer, Eds.) Wiley, New York, 1978, p. 359.

11   G. J. van Gurp. In *Semiconductor Silicon 1977.* (H. R. Huff and E. Sirtl, Eds.) The Electrochemical Society, Princeton, 1977, p. 342.

12   T. Mochizuki, K. Shibata, T. Inoue, and K. Ohuchi. *Jap. J. Appl. Phys.* **17** Suppl. 17–1, 37, 1978.

13   A. L. Crowder and S. Zirinsky. *IEEE J. of Solid State Circuits.* **SC-14**, 291, 1979.

14   S. P. Murarka. In *IEEE-IEDM 1979, Technical Digest.* IEEE, New York, 1979, p. 454.

15   S. P. Murarka. *J. Vac. Sci. Techn.* **17**, 755, 1980.

16   S. P. Murarka. *Silicides for VLSI Applications.* Academic Press, New York, 1983.

17   G. Ottaviani. *J. Vac. Sci. Techn.* **16**, 1112, 1979.

18   S. Iwamatsu and M. Ogawa. Presented at 157th Electrochemical Society Meeting, 11–16 May 1980, St. Louis, Abstract no. 174.

19   D. M. Brown, W. E. Engeler, M. Garfinkel, and P. V. Gray. *Solid State Electronics.* **11**, 1105, 1968; also in *J. Electrochemical Soc.* **115**, 874, 1968.

20   P. L. Shah. *IEEE Trans. Elect. Dev.* **ED26**, 632, 1979.

21   S. P. Murarka. *Appl. Phys. Lett.* **45**, 394, 1984.

22   G. Bomchil, G. Goeltz, and J. Torres. *Thin Solid Films.* **140**, 59, 1986.

23   Y. Sorimachi, H. Ishiwara, H. Yamamoto, and S. Furukawa. *Jap. J. Appl. Phys.* **21**, 752, 1982.

24   E. H. Rhoderick. *Metal–Semiconductor Contacts.* Clarendon, Oxford, 1978.

25   P. E. Schmid, P. S. Ho, and T. Y. Tan. *J. Vac. Sci. Technol.* **20**, 688, 1982.

26   M. Iwami, K. Okuno, S. Kamei, T. Ito, and A. Hiraki. *J. Electrochem. Soc.* **127**, 1542, 1980.

27   P. J. Grunthaner, F. J. Grunthaner, and A. Madhukar. *J. Vac. Sci. Technol.* **20**, 680, 1982.

28   H. Foll, P. S. Ho, and K. N. Tu. *Philos. Mag.* **A45**, 31, 1982.

29 S. P. Murarka, D. B. Fraser, A. K. Sinha, H. J. Levinstein, E. J. Lloyd, R. Liu, D. S. Williams, and S. J. Hillenius. *IEEE Trans. Electron Devices.* **ED-34**, 2108, 1987.

30 S. P. Murarka. *Thin Solid Films.* **140**, 35, 1986.

31 S. H. Ko, S. P. Murarka, and A. R. Sitaram. *J. Appl. Phys.* **71**, 5892, 1992.

32 A. R. Sitaram. Ph.D. dissertation. Rensselaer Polytechnic Institute, Troy, NY, 1990.

33 A. Appelbaum, R. V. Knoell, and S. P. Murarka. *J. Appl. Phys.* **57**, 1880, 1985.

34 M. Farooq. Ph.D. dissertation. Rensselaer Polytechnic Institute, Troy, NY, 1988.

35 W.-K. Chu, J. W. Mayer, and M. A. Nicolet. *Backscattering Spectrometry.* Academic Press, New York, 1978.

36 F. A. Baiocchi, N. Lifshitz, T. T. Sheng, and S. P. Murarka. *J. Appl. Phys.* **64**, 6490, 1988.

37 S. P. Murarka and S. Vaidya. *J. Appl. Phys.* **56**, 3404, 1984.

38 S. P. Murarka, E. Kinsbron, D. B. Fraser, J. M. Andrews, and E. J. Lloyd. *J. Appl. Phys.* **54**, 6943, 1983.

39 S. P. Murarka and D. B. Fraser. *J. Appl. Phys.* **51**, 350, 1980.

40 C. Canali, F. Catellani, M. Prudenziani, N. H. Wadlin, and C. A. Evans. *Appl. Phys. Lett.* **31**, 43, 1977.

41 S. Vaidya, S. P. Murarka, and T. T Sheng. *J. Appl. Phys.* **58**, 971, 1985.

42 S. S. Iyer, C.-Y. Ting, and P. M. Fryer. In *VLSI Science and Technology, 1984.* (K. E. Bean and G. A. Rozgonyi, Eds.) The Electrochemical Society, Pennington, NJ, 1984, p. 381.

43 J. J. Santiago, C. S. Wie, and J. Van der Spiegel. *Mats. Lett.* **2**, 477, 1984.

44 S. P. Murarka, M. H. Read, and C. C. Chang. *J. Appl. Phys.* **52**, 7450, 1981.

45 C. R. Tellier and J. Tosser. *Size Effects in Thin Films.* Elsevier, Amsterdam, 1982.

46 R. B. Marcus and T. T. Sheng. *Transmission Electron Microscopy of Silicon VLSI Circuits and Structures.* Wiley, New York, 1983.

47 P. J. Grunthaner, F. J. Grunthaner, A. Madhukar, and J. W. Mayer. *J. Vac. Sci. Techn.* **19**, 649, 1981.

48 R. J. Nemanich, R. T. Fulks, B. L. Stafford, and H. A. Van der Plas. *Appl. Phys. Lett.* **46**, 670, 1985.

49 W. J. Kaiser and L. D. *Bell. Phys. Rev. Lett.* **60**, 1406, 1988.

50 L. J. Schowalter and E. Y. Lee. *Phys. Rev. B.* **43**, 9308, 1991.

51 R.-T. Tung, S. Nakahara, and T. Boone. *Appl. Phys. Lett.* **46**, 895, 1985.

52 R.-T. Tung, J. C. Bean, J. M. Poate, J. M. Gibson, and D. C. Jacobson. *Appl. Phys. Lett.* **40**, 684, 1982.

53 L. J. Schowalter and E.Y. Lee. *Phys. Rev.* **B43**, 9308, 1991.

54 J. Wollschlager, J. Falta, and M. Henzler. *Appl. Phys.* **A50**, 57, 1990.

55 M. Farooq, S. P. Murarka, C. C. Chang, and F. A. Baiocchi. *J. Appl. Phy.* **65**, 3017, 1989.

56 S. P. Murarka, D. B. Fraser, A. K. Sinha, and H. J. Levinstein. *IEEE Trans. Electron. Dev.* **ED-27**, 1409, 1980.

57 S. P. Murarka and C. C. Chang. *Appl. Phys. Lett.* **37**, 639, 1980.

58 S. P. Murarka. *In Semiconductor Silicon 1981.* The Electrochemical Society, Pennington, NJ, 1981, p. 551.

59 W. Chu, A. S. Bhandia, and J. B. Stimmell. *J. Vac. Sci. Technol.* **B2**, 707, 1984.

60 H. Gant, H. Boetticher, and H. R. Deppe. *J. Vac. Sci. Technol.* **B3**, 1668, 1985.

61 R. Liu, S. P. Murarka, and P. Pelleg. *J. Appl. Phys.* **60**, 3335, 1986.

62 H. Jiang, C. S. Petersson, and M.-A. Nicolet. *Thin Solid Films.* **140**, 115, 1986.

63 S. P. Murarka and D. S. Williams. *J. Vac. Sci. Technol.* **B5**, 1674, 1987.

64 T. Hara, H. Takahashi, and S.-C. Chen. *J. Vac. Sci. Technol.* **B3**, 1664, 1985.

65 M. Ostling, C. S. Petersson, C. Chatfield, H. Norstrom, F. Runovac, R. Buchta, and P. Wiklund. *Thin Solid Films.* **110**, 281, 1983.

66 L. M. Williams. Appl. *Phys. Lett.* **46**, 41, 1985.

67 M. Y. Tsai, H. H. Chao, L. M. Ephrath, B. L. Crowder, A. Cramer, R. J. Bennett, C. J. Lucchese, and M. R. Wordeman. *J. Electrochem. Soc.* **128**, 2207, 1981.

68 P. Pan, N. Hsieh, H. J. Geipel, Jr., and G. J. Slusser. *J. Appl. Phys.* **53**, 3059, 1982.

69 S. Vaidya, T. F. Retajczyk, Jr., and R. V. Knoell. *J. Vac. Sci. Technol.* **B3**, 846, 1985.

70 J. Pelleg and S. P. Murarka. *J. Appl. Phys.* **54**, 1337, 1983.

71 V. L. Teal and S. P. Murarka. *J. Appl. Phys.* **61**, 5038, 1987.

72 A. R. Sitaram, J. Kalb, Jr., and S. P. Murarka. *Mat. Res. Soc. Symp. Proc.* **188**, 67, 1990.

73 M.-A. Nicolet. *Thin Solid Films.* **52**, 415, 1978.

74 M. Tanielian, R. Lajos, S. Blackstone, and D. Pramanik. *J. Electrochem. Soc.* **132**, 1456, 1985.

Aluminum- and Copper-Based Conductors

DAVID FANGER and ROGER TONNEMAN

Contents

4.1 Introduction

Aluminum and its alloys are currently the most widely used interconnect material for silicon-based integrated circuits (ICs). This chapter describes the techniques used to deposit interconnect metallization as well as some of the problems associated with interconnect deposition and delineation. Whenever possible, examples of the usage of analytical techniques are given, along with the limitations of these techniques.

History

Initially, a wide range of metals were used as the metallization step for the early discrete devices and the low-density integrated devices. Aluminum, gold, and other noble metal systems have all been used at one time in this stage of the technology. Aluminum has emerged as the interconnect metallization of choice for silicon structures due to a number of factors, primarily its ease of deposition, low cost, excellent adherence to silicon and its oxides, and its high-etch selectivity compared to silicon oxides.[1]

The noble metals are still used in some applications where the primary concern is the high conductivity of the interconnect. Structures such as unusually high-mobility transistors or solid-state lasers (fabricated from various III–V materials) require high-conductivity interconects. In these applications, since oxides of silicon are not present, the resistivity requirements outweigh the cost, etchability, stability, or other factors already noted.

Despite aluminum's widespread use, it is not without problems. The primary problem is that aluminum rapidly interdiffuses with silicon at temperatures of 400 °C or greater.[2] Since aluminum is essentially a p-type dopant, this interdiffusion creates rectifying contacts in silicon that are doped n-type. In addition, the aluminum can actually create a diffusion pipe through a diffused contact and cause shorts. Called "contact spiking," this phenomenon is discussed later in this chapter. From the aluminum–silicon phase diagram it is apparent that, by adding a few percent of silicon to the aluminum metal, the solubility of silicon in aluminum will be satisfied and contact spiking problems will be diminished. Thus, later silicon structures used aluminum with 1–2% silicon as the interconnect metallization. The amount of silicon to be alloyed in the aluminum metal is determined by the highest temperature seen after deposition. The lowest silicon concentration required to prevent spiking was determined from the phase diagram to be 0.8%.

With the shrinking size of new IC designs, metallization line widths also decrease. This means that the current density through any given area of metal lines has increased. This increased current density will actually cause mass movement in aluminum and its alloys. Such mass movement under an applied current is called electromigration. If aluminum is alloyed with a few percent of copper, the resulting alloy's electromigration resistance is superior to an aluminum–silicon alloy.[3] Hence, for many higher current applications such as Bipolar devices, aluminum–silicon–copper metallization is used instead of aluminum–silicon metallization.

Despite the resolution of the contact spiking problems and electromigration improvements, there are still two main problems with using aluminum-based alloys in today's devices: (1) hillock formation and (2) the need for improved electromigration resistance as line widths shrink even further. Hillocks are defined as the mass pileup near grain boundaries due to some external or internal stress. As an example, hillocks will form as the result of relief from temperature-induced stress. The difference between hillocks and electromigration failures is that, in electromigration failures, mass is moved from one location to another in the metal line. This results in a depletion of aluminum in one location, causing the metal line to fail. Hillocks generally do not cause metal line failures because mass is added to a specific area. The hillocks cause interlayer shorts (see Figure 4.1). Using aluminum with copper reduces both of these problems but does not overcome another issue that can lead to electromigration problems—metal step coverage, which can cause thinner cross sections over steps than on planar surfaces. The amount of material deposited on the side walls of steps and compared to a planar surface is termed metal step coverage. If the metal line cross section is thinned out (relative to the plane metal thickness), this can lead to an increased current density over that particular spot. To improve metal step coverage, new techniques are being developed or are being used. These techniques include using a refractory barrier metallurgy,

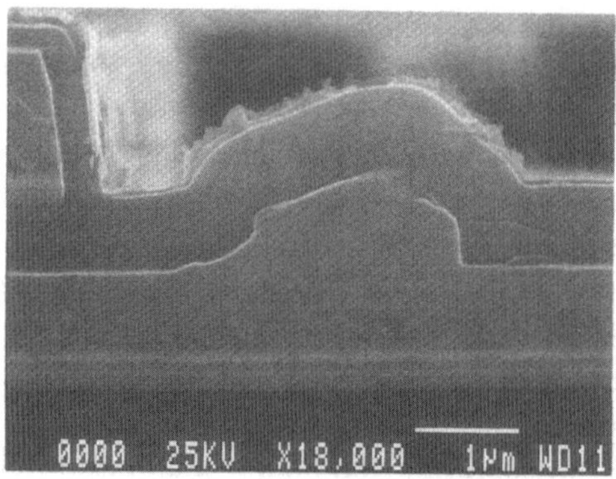

Figure 4.1 SEM photograph of a typical hillock in aluminum 2% silicon deposited at room temperature by planar magnetron sputtering.

aluminum with tungsten via fill or chemical vapor deposition (CVD) aluminum, bias sputtered deposition, deposition at elevated temperatures, and planarization of the underlying topography prior to metallization. Planarizing appears to be especially effective for multilevel metallization (more than one metal line). Tungsten deposition and its associated problems are discussed in a later chapter, as is barrier metallurgy. CVD aluminum is discussed later in this chapter.

For ULSI/GSI (ultra-large-scale integration/giga-scale integration) technologies, the most important requirement for metallization will be its resistivity and electromigration resistance. At these device-packing densities, typically greater than 4 million transistors on a device, the metallization line widths will be submicron. Although the search for the best conductor is ongoing, a leading contender appears to be pure copper. Researchers[4] working with this material report at least a factor of 10 in improvement electromigration resistance over aluminum–silicon alloys. Copper has a bulk resistivity of about 1.7 $\mu\Omega\cdot$cm versus aluminum's 2.7 $\mu\Omega\cdot$cm. Copper does have disadvantages[5] in that it does not adhere well to silicon dioxide, it is difficult to etch, and it corrodes easily in air.

4.2 Film Deposition

Techniques

This section briefly overviews some of the major techniques used for deposition of aluminum and its alloys on silicon substrates. More thorough discussion of these techniques is given in References 6 and 7, but brief discussion here will make the reader aware of some of the major limitations in each technique. Most of this section describes the most commonly used deposition technique, sputtering.

Any deposition technique must deposit the interconnect film uniformly over the substrate, deposit an aluminum alloy, and introduce no contamination during the deposition process. For a technique to be commercially viable, the following constraints must be satisfied also: the deposition system must be fairly easy to maintain, it must not be prohibitively expensive, and it must have a very high throughput of substrates.

Currently, the most common technique used to deposit aluminum is some form of physical vapor deposition (PVD). PVD includes evaporation, some form of sputtering, direct current (dc) or radio frequency (rf), be it diode or magnetron, and ionized cluster beam. Conductors can easily be deposited using dc sputtering. Dielectric deposition usually requires rf sputtering.

Evaporation was the first method to be used in aluminum deposition for IC applications because it was relatively inexpensive and quick and the equipment was easy to operate. For small silicon substrates and low-scale integration technologies, it satisfied all of the requirements noted previously. In evaporation, a source of aluminum is excited by some means so that the aluminum is vaporized. The vapor atoms will be deposited on any surface within the line of sight of the source. To produce the aluminum vapor, a variety of techniques have been used. The simplest is to coat a tungsten wire with aluminum and pass a current through the wire. As the wire heats up, the aluminum atoms are boiled off. This is simple, but it does not result in a very pure aluminum film since some of the wire material is mixed with the aluminum atom flux. An improvement is to direct a beam of electrons onto a crucible containing aluminum pellets. The electrons will heat up the aluminum in the crucible and the resultant atomic flux deposits on any sample within range. Use of an electron beam source to create the aluminum vapor is simply called "e beam evaporation." To reduce the incorporation of unwanted atoms in the condensing film, one typically performs all evaporations under vacuum in the ranges of 1.2×10^{-2} to 1.2×10^{-3} Pa. Since the aluminum atoms are distributed in nearly a cosine distribution, it is sometimes necessary to move the substrate under the atom flux to achieve good film uniformity.

As interconnect materials became more complex (the films became thicker and aluminum alloys started being used to prevent contact spiking), evaporation techniques proved to have many inherent problems. For example, the alloy compositions have different vapor phase rates, so the deposited film composition was hard to control, step coverage and thickness uniformity was very poor over larger silicon substrates, and the contamination level of most evaporators was undesirable. An improved PVD technique involving sputtering was, therefore, developed. Sputtering is defined as a technique that removes atoms from a surface by ion bombardment. There are many different ways to sputter any material. Sputtering is a physical process that in its simplest terms is analogous to knocking billiard balls off a rack of balls. This is just momentum transfer from an impinging argon atom to the aluminum atom. The sputtering process can be broken down into four steps: (1) a source of ions is created; (2) these ions are directed at and bombard a target; (3) atoms are ejected (sputtered) off the target; and (4) the sputtered atoms are transported to some substrate and condense on this substrate to form a film. All currently available sputter techniques involve using a glow discharge as a source of ions.

A glow discharge is a self-sustained plasma. The majority of commercially available metal-sputter systems use a form of diode-sputtering called planar magnetron sputtering. In this type of sputtering, a magnetic field is applied to the target to trap exiting electrons near the target surface. If the magnetic field is designed properly incoming atoms are confined to particular regions of the target. This serves two purposes. One, it increases the number of ions bombarding the target and increases the amount of material being sputtered from the target. The result is a very high sputter rate. Second, electrons will be ejected from the target along with atoms. These electrons can reach the substrate and cause undesirable heating of the substrate. The magnetic field will deflect these electrons, thus preventing this. Note, however, that the substrate is still heated during the deposition simply by the heat of formation of the growing aluminum film. Deposition rates are quite high, typically 150–200 Å/s. Also, sputtering systems have very low base pressures (compared to evaporators), so the mean free path of gases is long, which means that there will be fewer collisions between aluminum atoms and background gases. This means that there will be fewer unwanted atoms incorporated into the condensing aluminum film. Current generation sputter systems typically reach a base pressure of about 10^{-6} Pa, with sputtering occurring under a partial pressure of argon of about 10^{-1} Pa, typically. The major drawback of current sputter systems (relative to evaporators) is that they cost between \$1 and \$2 million. However, they currently are the only feasible way to deposit aluminum alloys (or multiple films) with consistent film properties and thickness control, good step coverage, and high substrate throughput (a typical sputter system can deposit a 1-μm film on greater than twenty-five 150-mm silicon substrates in 1 h).

Focused ion beam deposition is similar to planar magnetron sputtering except the source of the ions is different. In this technique, the ion beam provides the energy to knock electrons off the target. This technique has the possibility that deposition and patterning may occur in one operation. Since an ion beam is charged, it can be directed across the face of a target. This in turn will create a path of atoms that can be directed at a substrate. Ion beam deposition or ionized cluster beam deposition is being used in wafer-scale research applications.

The last technique discussed here is aluminum deposition utilizing CVD. In CVD aluminum, the film is produced as a result of the reduction reaction of two or three gases containing aluminum compounds. The aluminum is then transported to a substrate, where it condenses on the surface to form a film. Since the typical gases used in this reaction are organic-containing metallics, this technique is usually referred to as metallic-organic chemical vapor deposition (MOCVD). The one major advantage of MOCVD techniques over sputtering is that the film step coverage is superior. The primary reactant gases used to date have been trimethylaluminum (TMAl) and triisobutylaluminum (TIBAl). These gases have fairly low vapor pressure and TIBAl is unstable at room temperature. Therefore, the gases are hard to control. It has been noted that during deposition pure aluminum films can not be obtained due to carbon incorporation.[8] Newer techniques such as

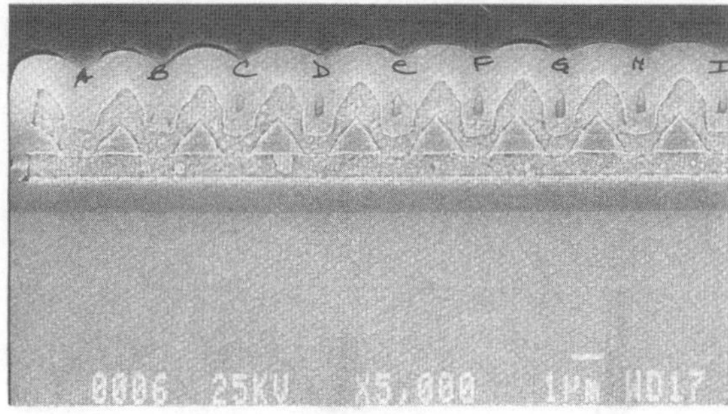

Figure 4.2   SEM photo showing step coverage degradation of a metal line due to shadowing. Metal is aluminum 2% silicon deposited at room temperature by planar magnetron sputtering. It covers the dielectric ridges (the dark triangles). The atom beam came in at an oblique angle from the left and the metal thickness reduction caused by shadowing can be seen in the troughs between the ridges.

using laser-assisted pyrolysis on these compounds have been tried[8] with limited success. However, MOCVD of aluminum is still an experimental technique.

Problems with Deposition

As mention previously, electromigration concerns are responsible for the development of improved interconnect materials. These concerns are also driving improvement in the depostion techniques. During aluminum deposition, the key parameters which affect electromigration are level of background gases incorporated (and what specie), grain size, film thickness/uniformity, and step coverage.

Step coverage and thickness uniformity are closely related. Within sputtering, the control of these parameters is through the geometrical considerations as well as the type of deposition. Sputtering was an improvement over evaporation due to improvements in deposition rate and target size. In sputtering, the field strength and geometrical arrangement of the magnets which define the confining field are key to giving good film uniformity and improved step coverage. In covering large substrates (150 mm and greater), there is a problem known as "shadowing." Shadowing is defined as the change in film thickness near a feature due to the angle of incidence of the incoming atom. Just as a tall building will create a shadow on the side opposite of which the sun strikes it, a step on the substrate will cause a reduction in film thickness (or a complete blockage) on the side opposite the incoming atom. See Figure 4.2 for an explanation of characterization by scanning electron microscopy (SEM) of shadowing.

Researchers[9] have found the sputtering ambient will have a profound impact on electromigration resistance. The most detrimental gases appear to be hydrogen, nitrogen, and oxygen. These contaminants will become incorporated into the

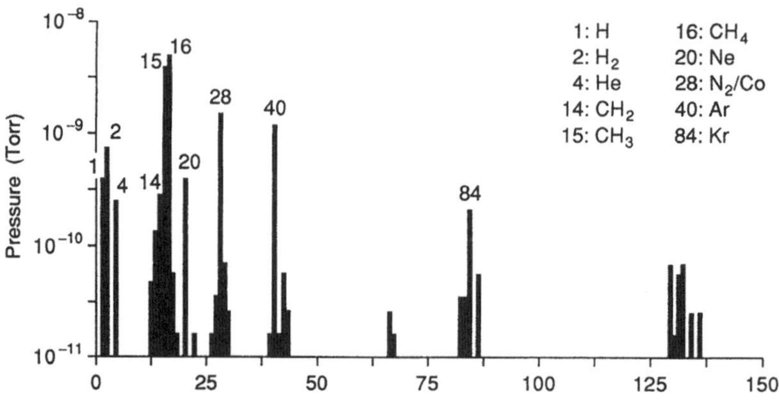

Figure 4.3   Typical RGA spectrum of a film deposition chamber at 1×10^{-6} Pa base pressure.

growing aluminum film (if their partial pressure is high enough). They will diffuse to grain boundaries in the grown film, reducing the effectiveness of late sintering operations. The result is that a film with a large percentage of contaminants will have smaller grain and a rougher surface than films with a lower contamination level. The number and size of grains in the aluminum film affect its electromigration resistance.[†] To reduce the impact of these gases and their effect on the resultant grain boundaries, there must be low partial pressures of these gases during deposition. In practice, the partial pressures of these gases can be reduced by a factor of 1000 or greater during deposition by gettering the ambient just prior to deposition. Another technique is to simply reduce the overall base vacuum pressure prior to deposition and utilize load locks to separate the deposition chambers from the atmosphere. The analytical tool to detect residual partial pressures is the residual gas analyzer (RGA). State-of-the-art RGAs will detect partial pressures over a large dynamic range. These tools can be used from base pressures (10^{-6}Pa) to deposition pressures (0.1 Pa). Figure 4.3 shows an RGA spectrum of a typical vacuum chamber prior to deposition and the effect of gettering. Residual gas incorporation can be detected in the grown film using secondary ion mass spectroscopy (SIMS). However, in production applications, it is best to monitor gas presence during the deposition or just prior to deposition so that corrective action can be taken to eliminate the sources of the residual gases. Since the RGA can provide real-time analysis of vacuum conditions, this is the analytical tool of choice during deposition, with SIMS the analytical technique for postanalysis of the films.

[†]See various papers in the Proceedings of the Fifth International Conference on Scanning Tunneling Microscopy/Spectroscopy and First International Conference on Nanometer Scale Science and Technology. *J. Vac Sci. Tech.*, **9** (2), March/April 1991.

Deposition techniques (amount of heat used or bias applied to the substrate) and incorporation of residual gases affect the grain size and morphology of the grown film. Nondestructive techniques operating on a macro level have been developed to monitor grain size and surface morphology. Although these techniques are easy to use and so are typically used in production environments, the resultant measurement is a combination of many factors. These techniques cannot indicate which single factor impacted grain size or morphology the most. These tools are simple gauges to indicate when the result is bad.

The first analytical tool is a noncontact technique to measure sheet resistivity, which is simply defined as the resistance of a square of material independent of its thickness. This noncontact technique utilizes an rf tank circuit to measure the voltage needed to null out the circuit after its capacitance has been changed. In practice, the film to be measured is placed within the air gap of a capacitor. The resulting voltage applied to null out the circuit can be calibrated with known standards and read as a resistance or using a fixed film bulk resistivity, a thickness. Although is is not as accurate as some other methods, we have shown that the film thickness as measured by this tool can be affected by the film grain size. Differences in apparent thickness as much as 5% due to grain size differences have been observed. Sheet resistance can also be measured using the four-point probe method, which, of course, requires contact with the surface. Another method is to use a reflectometer, which measures grain size and surface roughness. In typical applications a specific frequency of white light is directed normal to the surface of the film. The scattered light (dark field mode) or reflected light (light field mode) is detected by a photomultiplier, and the resultant output is calibrated as a percentage of reflected light. We have found that this technique is very sensitive to changes in grain structure due to temperature fluctuations, sputtered material, or residual gas.

Two additional interesting new techniques for measuring grain size and surface roughness are scanning tunneling microscopy (STM) and laser scatterometry (LS). The STM is proving to be a useful tool to investigate the surfaces of nanostructures and grain size[10] of various materials. Figure 4.4 gives an STM three-dimensional representation of an aluminum alloy surface at various deposition temperatures. Note how well-defined the grain structure is and how it changes with deposition temperature. In the second emerging technique, LS, laser light (typically helium neon) is directed at an angle normal to the substrate and the substrate is rotated about some angle perpendicular to the incoming light. The scattered light is detected by a movable photomultiplier. The resulting power spectrum (upper plot in Figure 4.5) is indicative of the surface roughness and grain size. This technique shows promise as a fast, nondestructive method to determine film surface properties. This technique may be able to replace optical reflectance for this purpose. Using a secondary technique (e.g., SEM) to calibrate the power spectrum, one can define a relationship to find surface roughness as a function of deposition condition. The lower plot in Figure 4.5 shows this calibration curve for grain size versus deposition

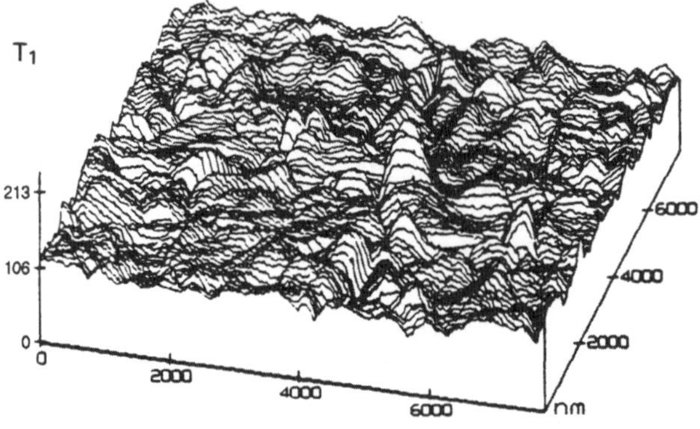

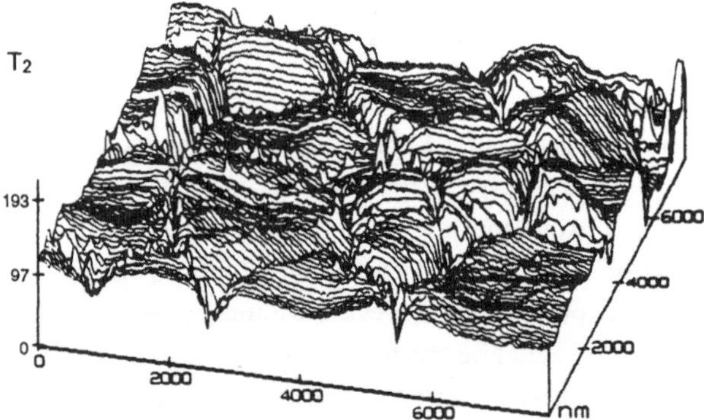

Figure 4.4   An STM of aluminum 1% silicon surface. Film deposited at two different temperatures, T_1 and T_2, by planar magnetron sputtering.

temperature. The LS technique utilizes a smaller wavelength excitation source than the reflectometer, thus allowing better resolution of surface morphology.

4.3 Film Growth

Substrate Surface Properties

Incoming substrate surface properties play a vital role in determining the final properties of the aluminum/copper conduction layer. Analysis of the incoming substrate surface is usually done when a problem in final film microstructure is detected, or when a new process is being characterized. An extensive analysis of substrate-film interaction effects should be completed in the early life of a semiconductor manufacturing process, and a high quality of incoming material should be maintained. In a semiconductor process, the "incoming" material to the metal deposition process will be incoming from a lithography/etch process for the formation of contacts or vias. Thus the substrate material will be an insulating film that

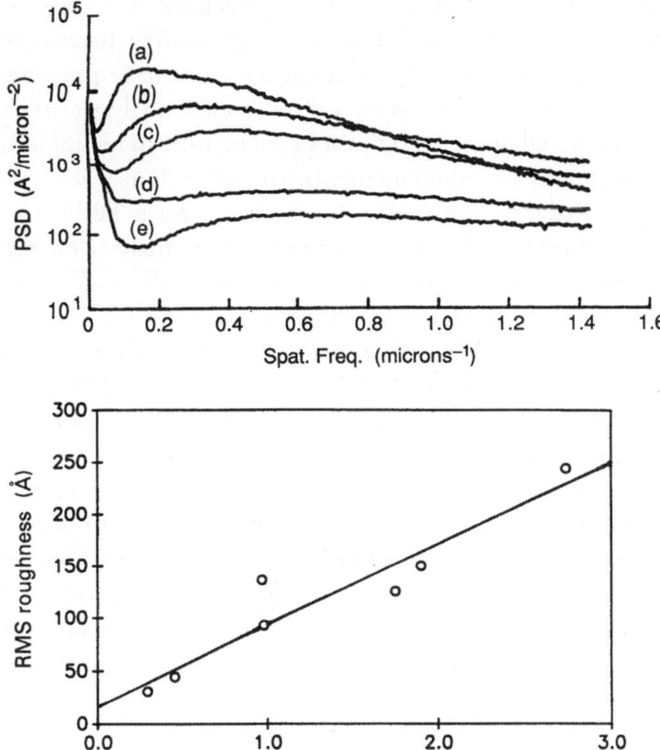

Figure 4.5   (*upper*) Power spectrum of an aluminum 2% silicon surface. Film deposited by planar mangetron sputtering under the following conditions: (a) 350 °C, rms = 150 Å; (b) 300 °C, rms = 131 Å; (c) 200 °C, rms = 97 Å; (d) 100 °C, rms = 44 Å; and (e) ambient, rms = 31 Å. (*lower*) Calibration curve made from independent SEM grain size analysis and power spectra at various temperatures.

has been exposed to organics. The insulating film can be a doped oxide (either phosphorus-, boron-, or phosphorus and boron-doped), silicon dioxide, silicon nitride, or silicon oxynitride. The organics may include photoresist, developers, various acids, plasma etch, or residual deionized water from a cleaning operation. Because of this processing the substrate surface may have particles, organic residues, acid residues, residual/native oxides, etc., that alter the physical properties of the conductive film to be deposited.

Contamination analysis of incoming substrates is done by the following surface techniques: Auger electron spectroscopy (AES), static SIMS, total internal reflection X-ray fluorescence (TXRF), and energy-dispersive X-ray spectroscopy (EDS). Each of the various surface analysis techniques may be the preferred technique for a specific application, depending on the issue needing resolution. The lead volume in this series, *Encyclopedia of Materials Characterization*, contains an in-depth discussion of each of these techniques and their relative strengths. AES may be utilized in an application such as via or contact contamination analysis because of the small probe size. Current one-micron and submicron semiconductor processes require submicron contacts and vias, and AES is the only elemental analysis surface technique able to analyze these small areas. Static SIMS may be used in an application where very low concentrations of contaminants (parts-per-million levels versus 0.1% for AES) are being analyzed, but the analysis surface area is fairly large (~100 μm). TXRF may be utilized for large-area contamination analysis on test wafers. TXRF cost is relatively inexpensive compared to SIMS. EDS may be utilized for the elemental analysis of high-level contaminants, but EDS is not a true surface analysis technique. The elemental detection limit is 0.1% within the volume sampled, and unless a thin window or windowless detector is utilized, EDS is not capable of the analysis of elements with an atomic number below sodium. Since it uses an e-beam, it is a small probe method, like AES.

Laser scattering analysis of the incoming substrate surfaces is the primary method of detecting surface particles. These particles may act as preferential nucleation sites in the subsequent metal deposition, or they may block the conduction path in the vias or contacts. The laser scatter particle analysis is not destructive, it is fast, and it is able to detect particles below 0.01 μm. Holographic laser analysis techniques are utilized for defined circuit particle analysis, but the region of analysis must have distinct repetition.

A heavily doped glass may have such a high concentration of dopant (for flow characteristics) that the surface adhesion of the conduction layer is poor. Corrosion due to high levels of phosphorous may also be an issue if the device is not enclosed in a hermetic passivation or package. Dynamic SIMS can be used to measure phosphorous/boron dopant concentration versus film depth. If the surface concentration is much higher than the bulk concentration, a surface treatment (wet etch) may be needed prior to metal deposition. AES depth profile has adequate elemental detection capabilities to differentiate phosphorus differences by depth at a concentration level of a few percent. A less accurate but simpler method is to utilize EDS with varying excitation energies. The higher the excitation energy, the deeper into the "bulk" the X rays are generated. This will only detect gross differences in the phosphorus concentration, and the depth resolution is poor (micron level). Care has to be taken to keep the X-ray generation within the dielectric film.

The interaction of the barrier layer metallization with the metal conduction layer is an important consideration. The surface mobility of aluminum and copper will be different on titanium, titanium nitride (TiN), titanium tungsten (TiW), and titanium oxynitride (TiON), for example, than on the various oxides. SIMS and TEM may be utilized for the characterization of the interaction of the conductor barrier layer. If the barrier layer is consumable (interacts with the conduction layer or silicon substrate), then the thickness and stoichiometry of the barrier layer should be fully characterized with Rutherford backscattering spectroscopy (RBS) until an optimum process window is established.

A later chapter discusses the material characterization of barrier layer metallizations in depth.

Surface Preparation

The formation process (nucleation and growth) of thin-film conduction layers plays a key role in determining the final properties of the deposited film. The nucleation and growth processes of PVD- and CVD-deposited aluminum or copper are similar in that atoms or molecules condense on the wafer surface, form islands, the islands grow larger, and the grains with preferential orientation grow faster than nonpreferentially orientated grains. Particles or contaminants on a surface can act as preferential nucleation and growth regions that "haze" the wafer. The nucleation site density can depend on the level of particles or contaminants on the incoming substrate. Contaminants can be incorporated into the conduction layer and alter the electrical properties of the film. Proper incoming surface preparation is needed in order to form a uniform, "clean" film. Figure 4.6 shows the interfacial region between silicon and aluminum as a result of incomplete cleans.

The primary concern with the incoming substrate surface is with the regions where electrical conduction will occur—the contacts and vias. The formation of a native oxide in the silicon contacts will always occur, and because the oxide will degrade the contact resistance, the native oxide must be removed. The removal of residual organics from the prior lithography or etch process is a major quality concern, so the substrate preclean should be able to remove residual organics. Wet

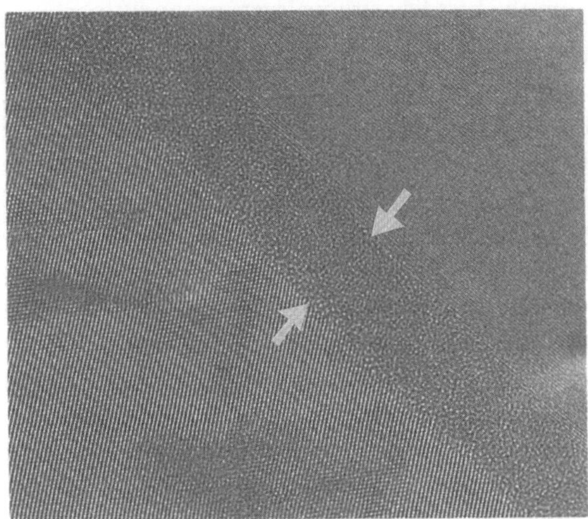

Figure 4.6   TEM analysis of a silicon–aluminum interface for a sample processed without a surface preparation prior to aluminum alloy planar magnetron sputter deposition. Areas between the arrows indicate the native oxide region. (Magnification ~1000 000×)

dip precleans are utilized with high success for removing organics and reducing the native oxide, but the subsequent rinse/dry operations and time delay (to a low vacuum) will always result in the reformation of the native oxide. Newer chemical vapor etching resolves the reformation of the native oxide because hydrogen will saturate the silicon surface-dangling bonds. Most new multichamber sputter systems have preclean stations built into the sputterer. These stations have low-pressure vacuums with high-temperature cleaning applications, sputter etch capability, and even reactive ion etch capability. An advantage of an in situ clean is that the wafer will be under a high vacuum until the metallization is deposited, and thus the formation of a native oxide will be minimized. A preclean process can be characterized by AES analysis of the contacts/vias, static SIMS, and also the end-of-line electrical characterization. This electrical characterization is completed on specialized test patterns that can consist of thousands of in-series connected contacts or vias.

Adhesion of the conductor film was a concern with older single-layer conductor technologies, but with new barrier layer metallizations, the adhesion of the conductor layer is usually not an issue. Dynamic SIMS analysis of a large surface before/after a preclean is a good tool for quantifying the added value of the preclean step. The analysis will usually be enhanced if a thin film of the normal conduction layer is deposited on the substrate after the preclean. This "cap" allows the atmospheric contamination of the surface during sample transfer to be negated, as the analysis region of interest is the interface of the substrate and conduction material. Static SIMS is a technique also utilized for surface analysis after a preclean. Here, no "cap" film is deposited, so contamination of the surface must be carefully controlled. In static SIMS, the etching rate is lowered to the point where a large fraction of the elemental/molecular species detected come from the first few atomic layers of the surface.

A dilute HF etch is usually used for the surface preparation and clean step before the first conductor layer. Various surfactant agents are utilized with the HF etch to enhance the "wetting" of the etchant with high-aspect-ratio contacts. A buffer is also usually added to the HF predeposition dip such that the etch rate will remain constant. The buffer most heavily utilized is ammonium fluoride. Ammonium fluoride can leave dendritic precipitates (observable optical or SEM analysis) if the rinse portion of the clean is not adequate. AES or dynamic SIMS depth profiles through the conduction film can readily resolve the high fluorine levels at the metal–oxide interface. Various wet chemical analysis techniques (atomic absorption, gas chromatography-mass spectrometry, inductively coupled plasma-mass spectroscopy) can be utilized for either the initial baseline or the later problem detection and isolation analysis. The subsequent metal layer predeposition clean cannot utilize the same HF mixture because the metal 1 conductor will be exposed during the etch—in situ rf plasma etches are the methods of choice for these cleaning steps.

Film Formation

The nucleation and growth process starts out with adsorption of the sputtered species, followed by migration and island growth. Defects on the incoming substrate

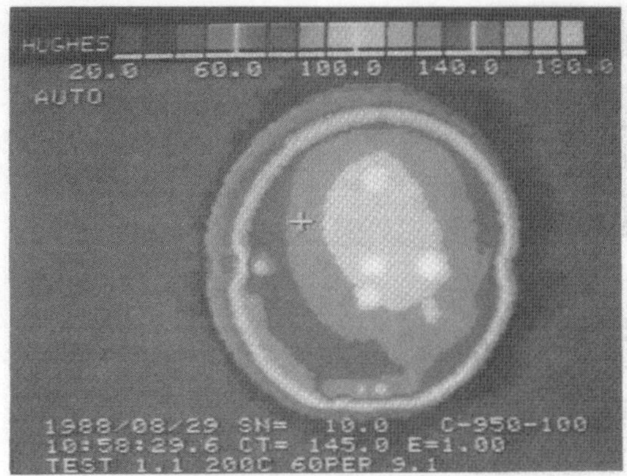

Figure 4.7   IR photograph of a wafer on a substrate holder with heat applied. Colors (seen here as shades of gray) indicate temperature. Temperature range is 20–180 °C.

(from either the preclean or the previous process step) can act as preferential nucleation sites. The surface energy of the sputtered species will affect the growth process due to the increase/decrease in surface mobility of the sputtered species. An increase in the cathode to substrate bias will cause higher surface mobility due to the increase in the sputtered species energy. Surface mobility is also largely dependant upon the interaction of the sputtered species with the substrate species. Higher density films result from films with higher surface mobilities—that is, the surface mobility allows vacancies to be filled. There are several ways to increase the surface mobility of the sputtered species, including higher cathode bias, back bias of the substrate, and additional heat from infrared (IR) heat lamps. One method of temperature characterization which is gaining acceptance in the semiconductor industry is IR photography, as shown in Figure 4.7. The uniformity of the heat transfer will directly affect the uniformity of the grain growth across the wafer and thus the across-wafer interconnect electrical resistances.

Island formation continues until the islands coalesce to form a continuous film. Larger islands with preferred lattice orientation grow at the expense of smaller islands. The grain growth is driven by the reduction in grain boundary energy and thus it will slow down when the grain size is equal to the thin film thickness. A fractal growth model simulates fairly closely the aluminum thin film growth. After the deposition process is complete, cross-sectional transmission electron microscopy (XTEM) shows the relationship between the large preferentially orientated grains and the small grains (Figure 4.8).

If enough thermal energy is supplied to the thin film during deposition, a problem can occur called grain boundary grooving. Electrical resistance of the conduction layer will increase due to the cross-sectional area of the metal lines being reduced. SEM, TEM, or STM analysis of the grain boundary should be completed if the as-deposited grain size is much thicker than the film thickness.

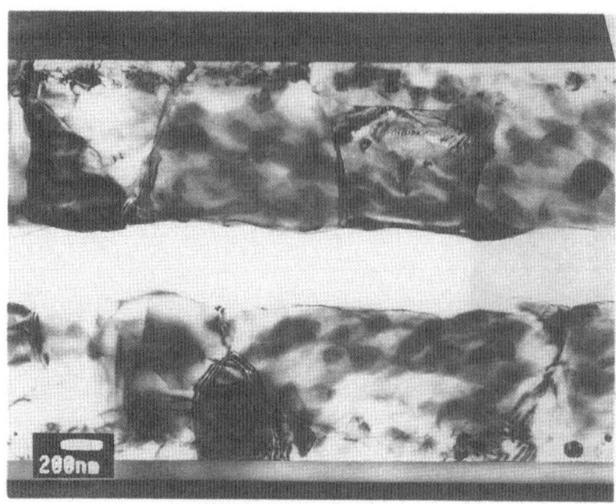

Figure 4.8   Cross sectional TEM showing relationship between the preferentially oriented grains and the small grains. Metal is aluminum 1% silicon deposited at room temperature by planar magnetron sputtering.

Microstructure

One of the primary modes of interconnect analysis is the study of the microstructure of the as-deposited layers. TEM is a fundamental analysis technique used for the characterization of either a planar film or in cross section (XTEM). The elemental X-ray analysis and the X-ray diffraction (XRD) capabilities of an analytical TEM instrument allow the analysis to go beyond obtaining a micrograph. Lattice orientation and elemental concentration analysis of individual grains, inclusions, precipitates, etc., will provide a well-rounded microstructure characterization. Scanning tunneling microscopy (STM) is used for surface morphology analysis along with LS and darkfield surface reflectance. SIMS is a primary elemental analysis technique of interconnect layers. Utilization of dynamic SIMS for elemental depth profiling analysis complements the use of XTEMs for problem analysis and new material characterization.

Patterning and Etching

The subsequent patterning and etching of the interconnect film alters the physical parameters of the deposited film. Grain growth occurs due to elevated temperatures during photoresist stabilization bakes, plasma etching, and plasma resist cleaning steps. The time/temperature effects on grain growth also depend upon alloy composition. TEM (grain size) and STM (surface morphology) are the primary analysis tools for the characterization of grain growth versus subsequent processing. Ancillary SIMS depth analysis of elemental composition versus thermal treatment is vital for a complete materials characterization.

Grain growth versus thermal process exposure can be quantified with TEM as shown in Figures 4.9 and 4.10. Each matrix in the figure had to be thinned to

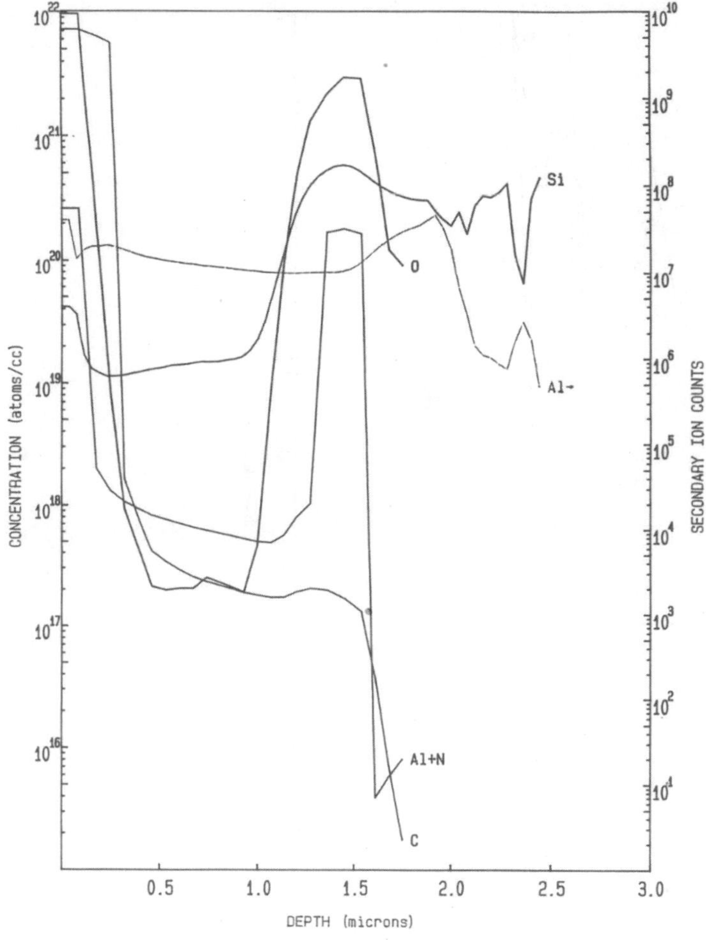

Figure 4.9 SIMS/TEM analysis of as-deposited aluminum alloy: (*upper*) SIMS depth profile and (*lower*) TEM micrograph.

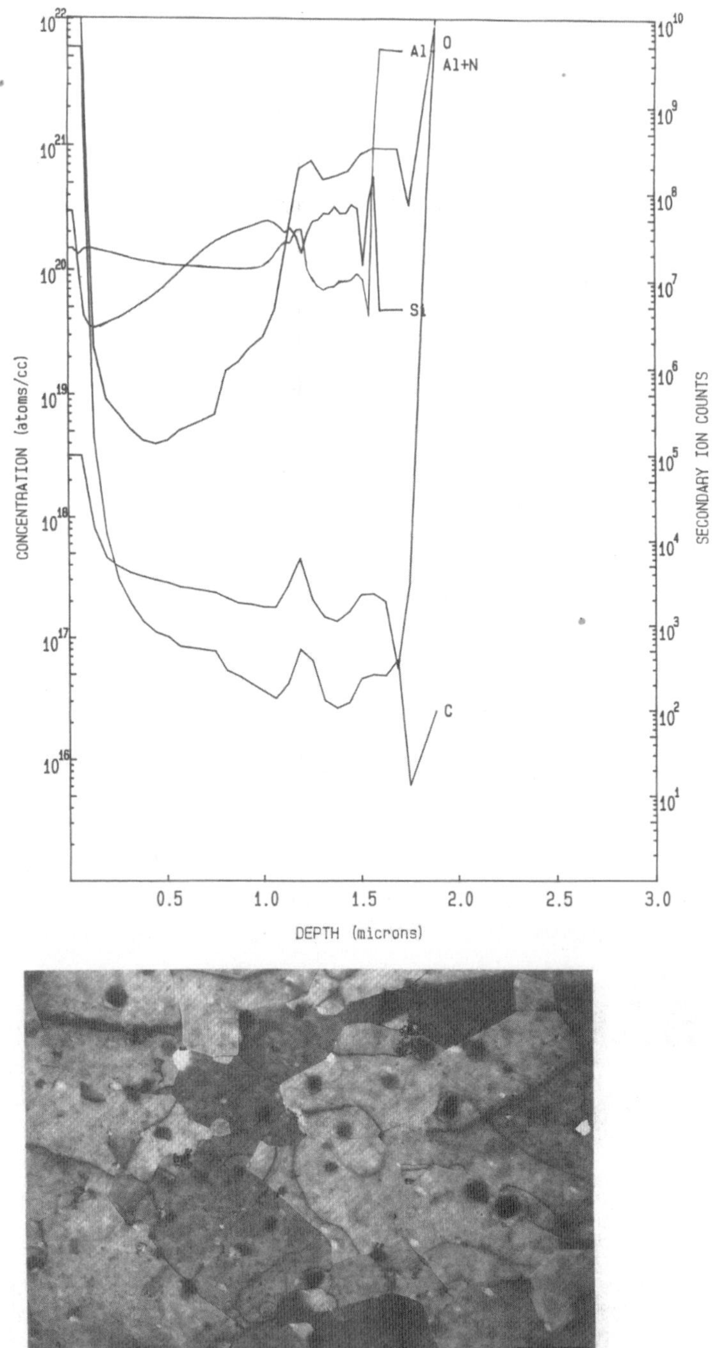

Figure 4.10   SIMS/TEM analysis of completely processed aluminum alloy: (*upper*) SIMS depth profile and (*lower*) TEM micrograph.

electron transparency thicknesses by grinding and etching through the backside of the wafer. The excellent contrast between grains is obtained because each grain has a lattice structure which is slightly misalinged with respect to the surrounding grains. Extended temperature treatments (as low as 255 °C) increase the mean grain size by an order of magnitude. Silicon precipitates are primarily present at grain boundaries and triple points, but can be seen within the grains. EDS of the TEM sample is utilized for elemental analysis of intergranular and intragranular micro-structure features. TEM diffraction patterning can be utilized for the characterization of lattice orientation for individual grains. If there is no hillock suppression layer on top of the conduction layer, hillocks will form from the stress-induced material transfer associated with the subsequent processing. One method of hillock suppression without an antireflective coating (ARC) is to "cap" the interconnect with a thin (300-Å) CVD oxide at room temperature, then ramp the wafers up to the final CVD temperature before the majority of the dielectric is deposited.

Several metal/lithography/etch interaction defects can occur that require materials characterization. Metal vertical and lateral hillocks can be formed by such interactions as well as through postdeposition heat cycles. Metal stringers are fine metal-to-metal shorts that result from either sidewall polymer lifting and blocking of the etch or from the movement of fine resist lines (to a point where they were not intended due to poor adhesion). Electrical characterization will pinpoint the physical location of the stringer failure and then after a onion peel (layer-by-layer strip back) the SEM, AES and EDS analysis can be completed on the stringer.

4.4 Encapsulation

Once the metal interconnect has been deposited, patterned, and delineated, it is necessary that the metal be protected from subsequent handling during further processing from the outside ambient to prevent corrosion. Materials used for this purpose are called encapsulating material. These materials are dielectric materials with the following properties: low pinhole density to prevent corrosion, good film coverage over the metal to also prevent corrosion, and high mechanical strength to protect the metal physically from scratches in further processing.

In addition to these film requirements, the deposition technique must be able to deposit an encapsulating film below 400 °C (typically). Above this temperature, the silicon in aluminum starts to diffuse so fast that metal film properties will change. Typical deposition techniques usually rely on plasma-enhanced chemical vapor deposition (PECVD). Typical encapsulating films are silicon nitride, silicon dioxide, silicon oxynitride, or a composite film of some combination of these films.

Although the encapsulating film is meant to protect the metallization, it can be detrimental to the metallization also. Any postmetallization temperature change generates a driving force that activates aluminum mass transport, creating voids in the metal line.[11] The difference in film stress between the overlying encapsulating material can cause hillocks or voids. If the encapsulating material is very compressive

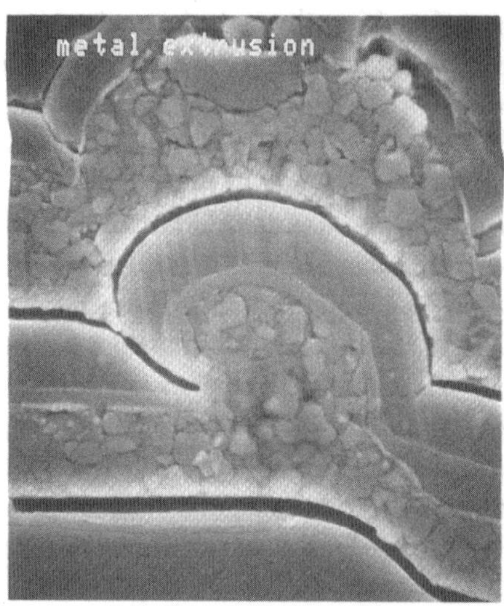

Figure 4.11 High-resolution SEM of a hillock formed in an aluminum 2% silicon metal line due to post metal deposition heat cycling. Deposition was at room temperature by planar magentron.

(as is PECVD-deposited silicon nitride), then hillocks may form. The encapsulating film can also lead to void formation. A study done by Yost et al.[12] indicates that the inherent stress in the metallization itself can lead to void formation over time even without an encapsulating film. Hillocks have been defined earlier in this chapter. Voids are simply the absence of metal in a particular area (typically along grain boundaries and triple points).

Analytical equipment used to study hillocks and voids formed by encapsulating material is typically limited to SEM and scanning transmission electron microscopy (STEM). Figure 4.11 shows a high-resolution SEM of a hillock in an aluminum metal line formed by annealing.

4.5 Reliability Concerns

Device reliability has come to encompass any functional failures over the planned lifetime of the device. Many of the issues discussed in this chapter, hillocks, voids, etc., cause device failure over time. Defects that are formed by mass migration (atomic movement) such as electromigration defects, stress voiding, or contact spiking impact device reliability. In actuality, any change in the metal line shape will change its current-carrying capabilities. These changes may impact the device's reliability in time. Therefore, planarization issues, line width control, metal thickess control all impact device functionality. This is so because all of these factors may change the patterned metal line profile. If the line profile decreases, then current density through this changed section of metal line will increase. As discussed earlier in this chapter, increased current density is the major driving force in causing electromigration failures.

In order to define the magnitude of the impact of a given defect on device failures, it is necessary to understand the propensity of a given defect to cause a failure. A model has been developed[13] which gives a mean time to failure (MTTF) of a device as a function of the current density and the defect's activation energy. This relationship is

$$MTTF = AJ^{-N}exp(E_a/kT)$$

where A is a proportionality constant based on the device geometry, J is the current density, N is some constant, E_a is the activation energy of the defect causing the failure, k is the Boltzman constant, and T is the absolute temperature. In the actual testing, N and E_a will be measured. The test for failures involves fabricating a special device with a simple metal pattern, packaging this device, and then placing these devices into a specially designed test board that will then be placed in an oven where the temperature can be varied over time for the duration of the test. During the test the device will be subjected to a given current. With a constant current, knowing the metal line characteristics, N can be calculated (it will generally be near 2). As the current is applied, the device's ambient temperature is raised. Failures are logged over time. Since these reliability devices are typically a simple metal line, the failure is usually an electrical open. The test is run over a long period of time (1000–2000 h). The resulting failures over time are plotted on a Arrhenius chart, and the slope will be the activation energy. Aluminum metallization failures have activation energies in the range of 0.4–0.8 eV.[14] By measuring devices with known defects against a control sample, one can determine activation energies for the defects.

It is usually necessary to analyze each failing device to determine what the cause of failure was so that an accurate appraisal of the failures activation energy can be made. Reliability testing typically generates 1–10 failed devices in each population tested, so analysis of individual failures is not impossible. The test devices are usually in some type of device package. Therefore, to analyze the failure one must open this package. If necessary, encapsulation layers or the metal layers themselves must be removed in order to identify the failure. Etches may be employed to remove the encapsulation, but these will generally affect the metal line and therefore may not be desireable. An improved technique commonly employs a commercially available focused ion beam (FIB) milling device to remove layers selectively. These tools employ an ion beam that can be focused and rastered across a section of the device. Typically, the commercially available ion mills also contain an imaging tool (SEM) so that the failure can be scanned before and after milling. It is really only through device testing (to find an activation energy) and post-testing failure analysis that activation energies can be assigned to a particular defect.

As discussed earlier, silicon interdiffusion and contaminants incorporated into the deposited film can lead to reliability issues. Table 4.1 gives the results of concentration measurements for the impurity levels for a thin-film aluminum alloy

Impurity Concentration, atoms/cm³

Sample	Oxygen	Nitrogen	Carbon	Silicon
V	2.3×10^{17}	6.5×10^{17}	2.5×10^{17}	1.4×10^{19}
W	9.0×10^{17}	7.0×10^{16}	1.6×10^{17}	9.0×10^{18}
X	2.5×10^{18}	1.0×10^{17}	1.0×10^{17}	1.1×10^{20}
Y	1.3×10^{18}	2.6×10^{17}	1.5×10^{17}	3.0×10^{19}
Z	4.0×10^{18}	2.7×10^{17}	8.0×10^{16}	1.2×10^{20}

Table 4.1   Impurity concentrations in an aluminum alloy thin film after various process steps. Sample V is as-deposited, W is after plasma etch X is after deposition, Y is after a bond-pad etch, and Z is after backside metallization.

after different processing steps, as determined by dynamic SIMS depth profiling. The example values given are at the midpoint of the film thickness, though, obviously, profiles over the whole thickness are available (see, for example, the SIMS depth profiles of Figures 4.9 and 4.10). The silicon distribution as a function of depth is modulated by each thermal treatment because the solubility of silicon in aluminum increases as the temperature increases.

Interconnect films utilizing aluminum usually contain silicon so that junction spiking will not occur (even when barrier metallization is utilized). The driving force behind junction spiking is an increase in solubility of silicon in aluminum at higher temperatures. Silicon will migrate from the lower interface to fulfill the solubility requirements, and aluminum will migrate to the vacated silicon region. If enough material is transferred, the aluminum will "spike" through the lower well interface, causing an electrical short.

Analysis of subsequent microstructure modulation can also be performed by electrical resistance measurement techniques. The electrical resistance is a function of grain size, film thickness, and impurity content. Usually the only "impurities" that are of sufficient concentration to modulate resistance are the alloying species of silicon, copper, titanium, etc. If a very poor vacuum system is utilized (due to contamination or leaks), the deposited film will have a high level of atmospheric or organic contaminatnts and dynamic SIMS analysis can pinpoint the elemental species.

TEM analysis of interconnect layer as a function of subsequent IC processing can also be utilized to quantity the type, number, and distribution of other crystallographic defects. Point defects (vacancies, interstitial), line defects (dislocations), and planar defects (stacking faults, twins, grain boundaries, and interfaces) can be quantified for each of the subsequent processing steps.

The interaction of grain size with lithography-defined line width is important because of the electromigration effects associated with triple points. A triple point is the thermodynamically stable 120° interaction between three grains. If the gain siz is smaller than the line width, the line resistance will be high and the electromigration

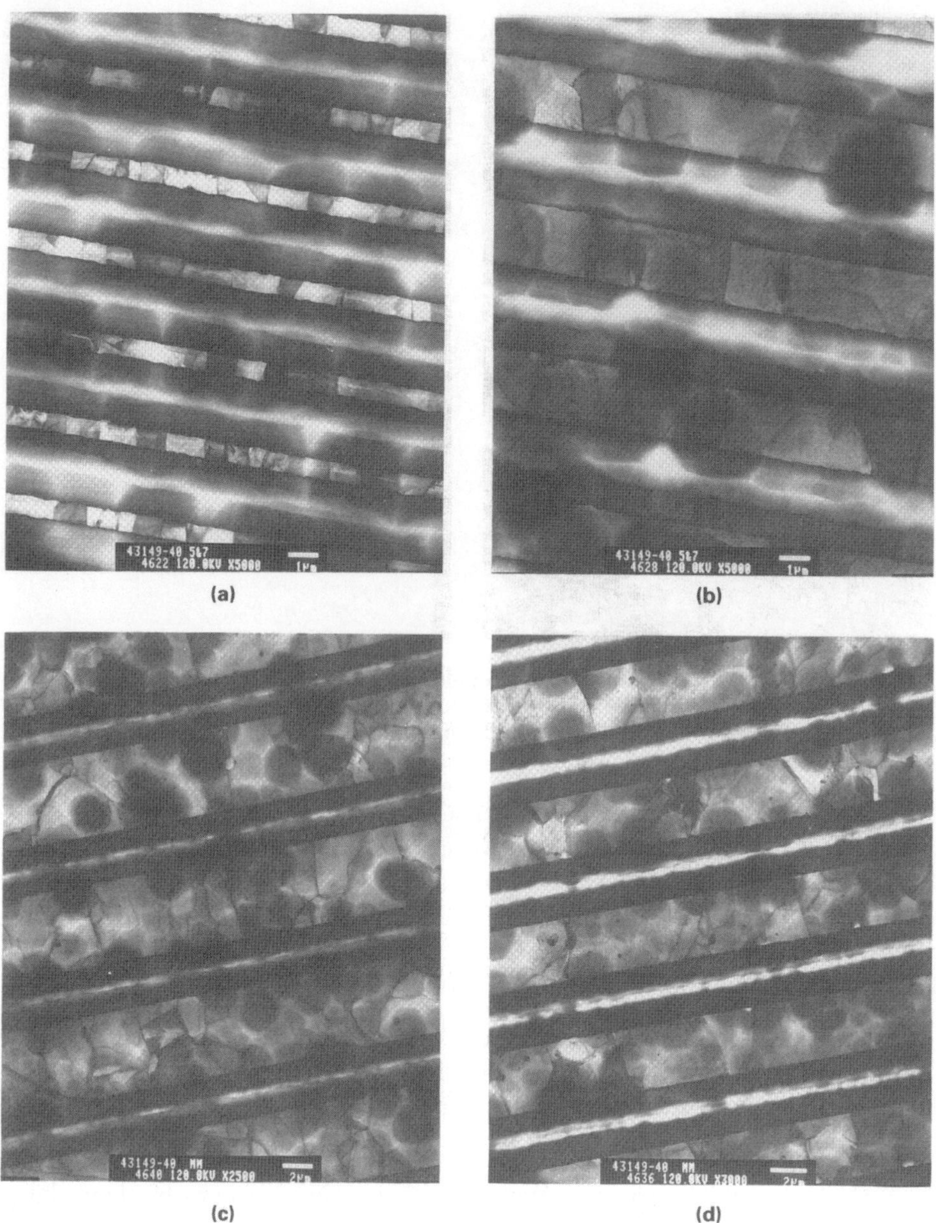

Figure 4.12 A series of TEMs showing grain size as a function of different line widths: (*a*) 3 µm, (*b*) 5 µm, (*c*) 7 µm, and (*d*) 9 µm.

MTTF will be low. If the grain size is larger than the line width, a bamboo structure will be present for the grain structure of the metal lines. Because of the lack of triple points, the bamboo structure will exhibit excellent electromigration MTTF and

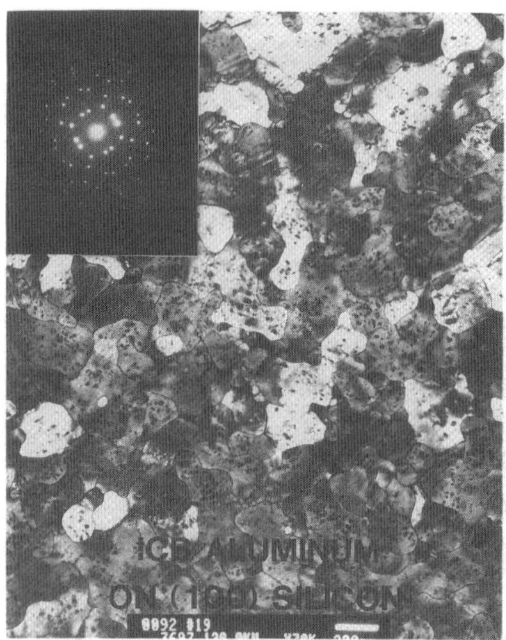

Figure 4.13   TEM micrograph of ICB Al on (100) silicon. Note unique grain boundary structure and lack of 120° triple points. Insert is small area electron diffraction pattern on sample.

lower line resistances. The effect of the grain size diameter versus the width of the metal line is shown in Figure 4.12. TEM is utilized here for the resolution of grain size on the etched lines, but SEM analysis after a reactive ion etch (RIE) highlight etch would also be able to delineate the grain structure. The thinner the metal line width, the smaller the grain size will be. If a bamboo structure is desired, the grain growth should occur before the definition of the lines, and the grain size should be maximized but not grooved.

Another deposition method utilized to minimize triple points is ion cluster beam (ICB) deposition of aluminum on (100) silicon,[15] as shown in Figure 4.13. The unique ICB grain structure contains no triple points.

Aluminum copper grains do not necessarily need to be of a bamboo structure because of the excellent electromigration resistance already shown by these alloys. The copper reacts with the aluminum to form $AlCu_3$ precipitates that can act as dislocation pileup zones and thus restrict material movement.

As the interconnection complexity increases, another cause for device failure is the effectiveness of the interconnect between the fabricated device and its package. Generally, the weakest point of this interconnection is the point where the package wire comes into contact with the metallization itself. In order for the device to be connected to the package, areas of large flat metal on the device perimeter are made. These areas are called bond pads since this is where the metallization and the

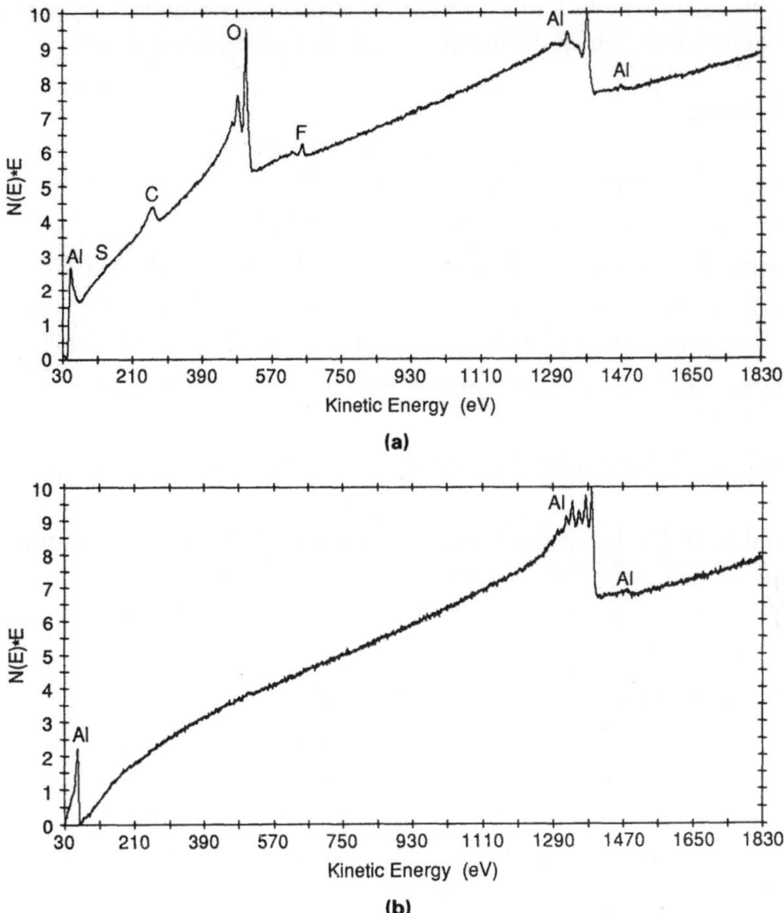

Figure 4.14   AES studies of a metal bond pad after etching of the encapsulating material: (a) as-received surface and (b) surface after cleaning of the as-received surface.

package wire will be bonded together. The actual bonding is done using ultrasonic or thermal techniques. The bond wire is typically aluminum with some silicon or gold. During the bonding process, the surface of the pad area will effect the resultant bond. Surface cleanliness is very important for good bonding to take place. Residual chemicals from the encapsulation etch or residual passivation material left on the pad will cause incomplete bonding and will lead to failures. In the presence of moisture, gold and aluminum will create an intermetallic compound (the so called purple plague) that weakens the bond and cause failures. The best tool to analyze the bond pad surface is AES. An AES spectrum of a representative area of an aluminum bond pad is shown in Figure 4.14*a*. Note the presence of fluorine and the excessive amount of oxygen, which are remnants of the passivation etch process.

Also note how these are removed after some surface cleaning via sputtering in the AES system (Figure 4.14*b*), which demonstrates that they were surface species.

References

1   A. J. Learn. *J Electrochem. Soc.* **133** (6), 394, 1976.

2   P. Burggraaf. *Semiconductor Internation.* 73, Nov. 1985.

3   J. K. Howard, J. F. White, and P. S. Ho. *J. Appl. Phys.* **49** (7), 4083, 1978.

4   J. S. Cho, H.-K. Kang, M. A. Beily, S. S. Wong, and C. H. Tong. SRC Pub. C91137. Semiconductor Research Corp., Triangle Park, NC, 1 Feb. 1991.

5   S. P. Murarka. SRC Pub. C90555. Semiconductor Research Corp., Triangle Park, NC, 31 Oct. 1990.

6   J. L. Vossen and W Kern. *Thin Film Processes.* Academic Press, 1978, chapts. 2 and 3.

7   S. Wolf and R. N. Tauber. *Silicion Processing for the VLSI Era*, Vol. 1. Lattice Press, Sunset Beach, CA, 1986, chapt. 10 and p. 375.

8   T Cacouris. SRC Technical Report T90133. Semiconductor Research Corp., Triangle Park, NC, Dec. 1990.

9   G. Quierolo et al. *J Vac Sci Tech.* **7** (3), 651, 1990.

10   P. B. Ghate. *Solid State Technology.* 113, March 1983.

11   J. G. Ryan et al. *J. Vac Sci. Tech.* **8** (3), 1474, 1990.

12   F. G. Yost, A. D. Romig, and R. J. Bourcier. "Stress Driven Diffusive Voiding of Aluminum Conductor Lines: A Model for Time-Dependent Failure." SAND88-0946, Sandia National Laboratories, Albuquerque, NM, 1988.

13   J. R. Black. *IEEE Tr. Electr. Dev.* **ED-16**, 338, 1969.

14   H. H. Hoang and J. M. McDavid. *Solid State Technology.* 121, Oct. 1987.

15   M. C. Madden. *Appl. Phys Lett.* **55** (11), 1077, 1989.

Tungsten-Based Conductors

ROC BLUMENTHAL and GREGORY C. SMITH

Contents

5.1 Applications for ULSI Processing

Tungsten is used in integrated electronics as a high-conductivity interconnect metallization, as a plug for vertical contacts and vias between metallization layers, or as a barrier between aluminum and silicon. Tungsten can be deposited by evaporation, but sputtering and chemical vapor deposition (CVD) are the preferred techniques. CVD films offer numerous advantages over sputtered films because of lower resistivity and superior conformality in small via holes. CVD also allows selective deposition of tungsten on metal and silicon. CVD tungsten has been fabricated from tungsten chloride, fluoride, and carbonyl. The reduction of the halide vapors has been accomplished in hydrogen, various polysilanes, germane, phosphine, and diborane. The most common CVD processes involve tungsten hexafluoride and either hydrogen or monosilane. This chapter focuses on analysis of chemical vapor deposited films and their peculiarities.

In integrated circuit (IC) processing, interconnections are fabricated by deposition of uniform layers of metal which are subsequently masked by patterns formed lithographically in photoresist. The metal is then etched to leave the interconnect wiring. This is connected with wiring layers below it through interleaved dielectric layers such as silicon dioxide using via holes. These are typically 1 μm or less in diameter (in the VLSI era). Current technology in the development laboratory makes use of vias with 0.4-μm diameter. Via holes range from about 0.5 to 2 μm in depth, depending upon the application. The main problem which CVD tungsten solves is one of step coverage, or the coverage of the metallization layer as it deposits onto the wall and bottom of the contact holes. Since sputtering and evaporation techniques are performed at low substrate temperatures, they suffer from poor step coverage due to the low surface mobility of the deposited species. Although modern sputtering techniques such as wafer heating and resputtering offer improvements, the conformality of such films in extremely small contacts is insufficient to meet the demands of manufacturing ICs with submicron features.

Chemical vapor deposited tungsten exhibits excellent step coverage in contact holes with depth-to-diameter ratios of up to two or more. The $WF_6 + 3H_2 \longrightarrow W + 6HF$ reaction is in wide industrial use for depositing blanket metal films for interconnect and barrier applications. These films are generally deposited on sputtered metal nucleation layers.

CVD tungsten is used either as part of a lateral interconnection layer or simply as a conductive plug for vias so that the higher conductivity sputtered metal does not have to carry current down into the via hole, where it would otherwise have been unacceptably thinned.

CVD tungsten layers can be used in several applications. They can be used as blanket barrier layers to prevent interaction between aluminum and the silicon at the bottom of contacts. Using either selective deposition directly, or an etchback process from a blanket layer, CVD tungsten can also be used as a conductive plug to fill contact holes. In addition, the metal can be used as a lateral interconnect layer by itself, if the high film resistivity (typically >8 $\mu\Omega\cdot$cm, more than twice that of sputtered aluminum alloys) can be tolerated. A primary advantage this approach offers is its superior reliability (with regard to electromigration, corrosion, and stress-induced voiding) compared to that of polycrystalline-sputtered aluminum or its alloys.

5.2 Deposition Principles

The deposition occurs by the reduction of a tungsten-containing compound, most typically tungsten hexafluoride or less often tungsten carbonyl, with a hydrogen-

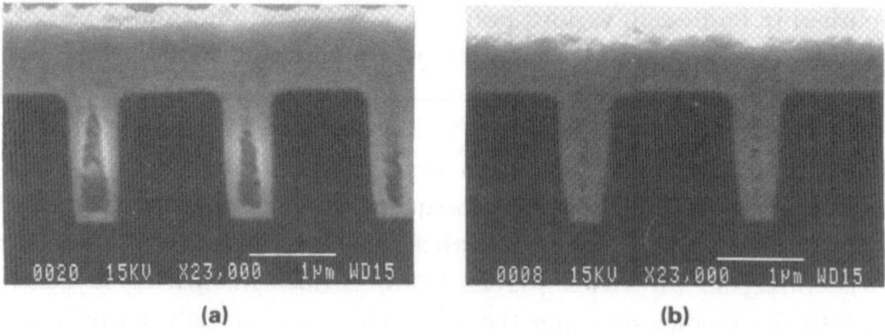

Figure 5.1   SEM cross section of (a) nonconformal and (b) conformal blanket tungsten film.

                                   TUNGSTEN-BASED CONDUCTORS   Chapter 5

containing compound, most typically molecular hydrogen or silane. The source gases will thermally decompose on heated surfaces (either the substrate or reactor surface) and react to form elemental tungsten and fluorinated byproducts (SiF_4 and H_2 for silane chemistry, HF for H_2 reduction reaction).

CVD of tungsten can be done in two modes: blanket and selective. The mode is primarily controlled by the substrate surface conditions, and the same deposition chemistries can be used for both modes. In addition, the reactant gas flows (partial pressures), reactor pressure, and temperature will determine the film properties of the deposition and will partially control the mode of deposition.

5.3 Blanket Tungsten Deposition

If a thin-metal layer is sputtered on top of an insulator layer and down into the contact holes, the tungsten layer will deposit nonselectively as a blanket layer over the entire surface and into the contacts. The contact filling capabilities of this process are not limited by contact depth variations but contact size, so it is a suitable method for filling small, deep contacts of different depths at the same time. Also, the deposition rate is not as sensitive to the type of substrate at the bottom of the contact since the substrate is covered by the thin sputtered metal film.

The two primary reaction chemistries are the H_2/WF_6 and the $SiH_4/H_2/WF_6$ systems. In general, due to the high deposition rate of the silane-based system, it is essentially transport limited throughout the range of deposition temperatures of interest (300–500 °C). Thin films deposited with this chemistry are typically used to protect the substrate or underlying sputtered adhesion layer from exposure to high partial pressures of WF_6 during a subsequent deposition of hydrogen-reduced blanket tungsten.

Film Thickness

Although thickness measurements may seem straightforward, in the case of CVD tungsten deposition, determining the film thickness is complicated by the roughness of the film. One method of measuring film thickness is using cross-sectional scanning electron microscopy (SEM) analysis, as in Figure 5.1. Typically, the underlying oxide is well-defined, but the outline of the thin sputtered metal adhesion layer is not well-defined, particularly if titanium-doped tungsten is the sticking layer. The sticking layer thickness can be estimated from analysis on test wafers prior to tungsten deposition. Defining the top of the film is the most difficult problem, due to its roughness. This means that thickness itself is a subjective parameter. Even if some roughness averaging technique is adopted for defining local thickness, many measurements will be necessary to obtain across-sample variations, which makes SEM too difficult for process control. Thickness can also be determined by measuring the mass gain of the sample during deposition. This technique can be used because of the high mass density of tungsten (~19.3 g/cm^3). The advantage of this technique is that it can be done quickly and simply without any special sample preparation. Mass gain measurements very sensitively determine W thickness (~30 Å per milligram of mass gain on a 150-mm wafer), but suffer from the need to assume a density for the film to convert the mass gain into thickness. Blanket tungsten films range from 70 to 99% dense, depending on thickness and deposition technique. Also, the thickness obtained is an average value for the entire sample and ignores any across-sample variations. This technique also requires that the deposition be limited to the surfaces of interest; if deposition occurs on the back or the edges of the sample, the mass gain value will reflect total deposition, which can cause overestimation of the thickness.

A third technique is to use a stylus to trace the film surface over a step etched in the film. This type of measurement does require special preparation of the film and is dependent on the roughness of the film. The stylus will tend to measure the tops of the grain peaks, depending on the size of the stylus tip. Also, the technique will not distinguish the sputtered sticking layer from the tungsten film. This technique is a localized measurement, and many measurements at different locations are needed to characterize across-sample uniformity. In general, film thicknesses will most likely be overestimated by this technique.

X-ray absorption and backscattering of beta radiation from a [109]Cd source have also been used to determine tungsten film thickness. Both techniques require calibration on the films in question because they measure the areal number density of tungsten atoms. Hence, in order to calculate the film thickness, one must know the density of the material. They do have the capability of making local film thickness measurements, however, which is an advantage over the mass gain technique.

Film Conformality

Film conformality is the most important film property for CVD blanket tungsten deposition, and the best analysis technique is cross-sectional SEM[1] (see Figure 5.1.) Specially prepared samples with trenches or holes of the appropriate size are prepared, and the sputtered sticking layer is deposited over the surface of the sample. CVD tungsten is deposited on the samples, and each is cleaved and examined under the SEM. Since the tungsten may be harder than the surrounding dielectric film, it is often difficult to cleave a sample so as to expose a void in the tungsten at the center of a small round contact hole. Seeing the void in the center of a tungsten contact plug usually requires careful mechanical polishing of the edge of the sample, along with a light hydrogen peroxide and dilute HF stain to remove polishing debris from the edge.

Film Resistivity

The resistivity of the film is another important parameter if the film itself is to be used as a primary conductor in microelectronic devices. This is a calculated parameter, which requires measurement of the film thickness and sheet resistance at a particular location. To obtain sheet resistances, two types of measurements can be done—a four-point-probe measurement which requires contacting the film surface or an eddy current measurement which can be done without contacting the surface. Both techniques are localized. The area tested is determined by the probe spacing in the case of the four-point probe. The readings overestimate the actual local sheet resistance by less than 3% when the probe is more than two probe spacings from the edge of the conducting film.[2] The wafer surface can be mapped by moving the probe to various locations. A single-sheet resistance reading, combined with a localized thickness measurement (as obtained by SEM or stylus) will give the resistivity of the film at that location. If mass gain is used for thickness measurement, an average value of sheet resistance taken from measurements over the entire sample can be used to calculate an average resistivity. Finally, if the resistivity of the film is constant over a sample, mapping the sheet resistance values over the sample will allow the determination of the thickness uniformity over the entire sample. This provides a quick and simple way to access film thickness uniformity without requiring numerous thickness measurements to be made. The sheet resistance mapping can be highly automated and fast, in contrast with stylus or SEM thickness determinations. Sheet resistance measurements can also be used to quantify barrier properties, as was done by Gutierrez et al.[3] to examine CVD W barrier properties between silicon and aluminum. They annealed aluminum films sputter-deposited upon 200 nm of CVD W on silicon in forming gas, nitrogen, and hydrogen at temperatures up to 550 °C. The sheet resistance of the aluminum films rose by a factor of four between the temperatures of 500 and 550 °C for the hydrogen and forming gas-annealed films. The nitrogen-annealed film rose by a factor of more than three. This was attributed to WAl_{12} formation, not to diffusion of silicon through the tungsten.

Film Stress

Film stress is a parameter that must be characterized and controlled to insure the integrity of the film and because of its indirect linkage to adhesion to the substrate. There are four techniques used for measurement of this parameter; most commonly used are capacitance and laser measurements of the wafer bow before and after deposition. The capacitance measurement is done over a small area of the sample, whereas the laser measurement is done along a particular diameter of the sample. The results of both techniques, combined with the elastic modulus and Poisson's ratio of the substrate and the film and sample thickness, yield a single stress value. These measurements must be made in several locations to determine if there is any asymmetry to the stress on the sample. A third technique using optical diffraction yields a wafer map showing stress variation in the form of "rings" on the image. The number of rings can be used to determine the stress value; this technique does give an image of the overall stress profile across the wafer. The fourth technique is X-ray Lang topography, in which the direction variation of the c-axis of the silicon crystal is measured. From this the radius of curvature and film stress are calculated.

Stresses in tungsten have been found to have two components—one which is interfacial and one which is a bulk property.[4] Therefore, the total film "stress" is dependent upon film thickness. The relative values of the bulk and interfacial stress depend upon the grain size in the tungsten and the type of sticking layer used for the blanket deposition. Total stress has the form

$$S = A/t + B$$

where S is total stress, A is interfacial stress, t is the thickness, and B is bulk stress. Therefore, in order for a tungsten deposition process to be characterized, the interfacial parameter and the bulk parameter must be measured, which requires measuring samples with at least two thicknesses from the same process.

Tensile stress in tungsten has also been found to be reduced by increasing the deposition temperature. This probably is related to improved nucleation of the first film to deposit.

Surface Roughness

As already mentioned, the roughness of the tungsten surface has a major effect not only on the metrology used to determine film properties, but also on the ability to integrate the film into a process flow for microelectronic applications. The roughness affects the patterning and etching of the film and can also affect depositions made on top of the film. SEM analysis can be used to observe the phenomena, but it is difficult to obtain an "average" roughness value for a surface area. The roughness can be indirectly measured by measuring the reflectivity of the surface for a given wavelength of light; the roughness of the surface will effect this measurement and can be correlated to these results. The measurement, which typically is done as percentage at a particular wavelength against a standard sample, is a localized measurement and can be used to map the surface of the wafer without requiring any special preparation. The weakness of this method is that there is not a one-to-one relationship between a measurement and a specific surface roughness; two difference roughnesses can have the same reflectivity values. This can only be resolved by SEMs of the surface to determine the type of surface asperities present.

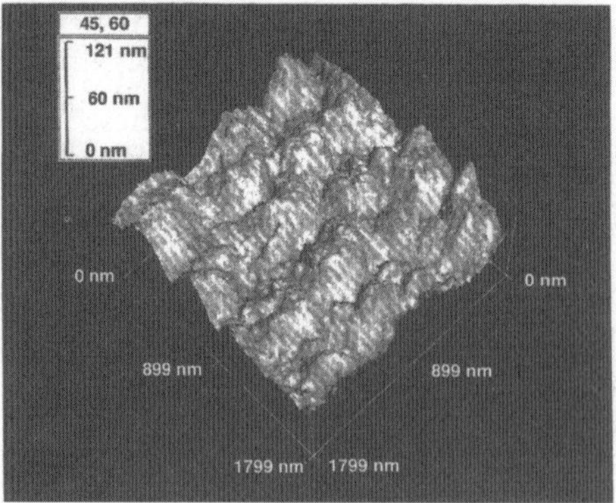

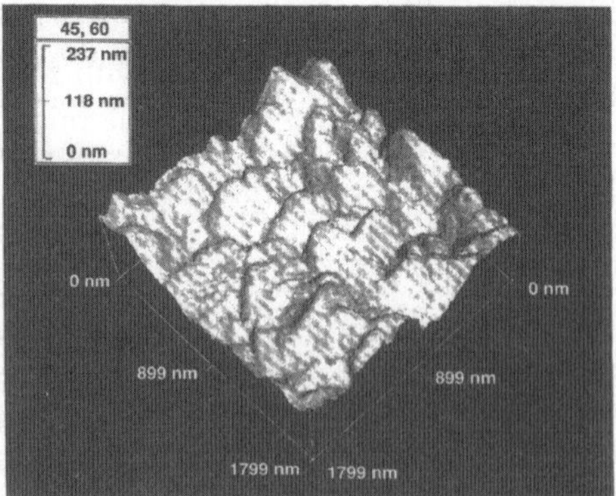

Figure 5.2   STM topographs of two different CVD blanket tungsten films.

The best technique is scanning tunneling microscopy (STM), which can map out the localized surface and produce images and quantitative roughness factors characteristic of a surface morphology (see Figure 5.2). The range and frequency of the feature sizes can be determined from the data and can be correlated back to reflectivity measurements. The area examined with this technique is small, so several measurements across the sample must be done to insure the correct representation of the surface is found. Figure 5.3 shows that the reflectivity of the films correlates directly to the deposition regime, that is, whether it is surface reaction-controlled or limited by the rate at which reactant can be supplied to or byproducts removed from the surface.

Film Microstructure

The film microstructure ultimately determines the fundamental properties of stress and resistivity. Analysis of the grain structure of CVD tungsten films is straightforward. Both plan-view SEM and TEM analysis can be used to examine grain size, as shown in Figure 5.4. The grain texture, or preferred orientation, can be found using standard X-ray diffraction (XRD) techniques for large-area samples, whereas selected area diffraction (SAD) done on TEM samples can provide information about local grain orientation. For cross-sectional information, SEM provides the most information; TEM cannot typically resolve cross-sectional grain structure since sample preparation is complicated by the hardness of the metal and the absorption of the tungsten is high enough to obscure any features. Figure 5.5 shows the effect of deposition temperature on the preferred orientation of the tungsten films. Changing deposition regimes also alters the value of this parameter.

5.4 Selective Tungsten Deposition

Selective deposition takes advantage of the fact that the chemical reaction itself can be inhibited on insulator surfaces due to the limitations of the reducing agent to dissociate while it occurs on conductor or semiconductor surfaces. When deposited in this mode, a tungsten film is grown on metal or semiconductor surfaces while

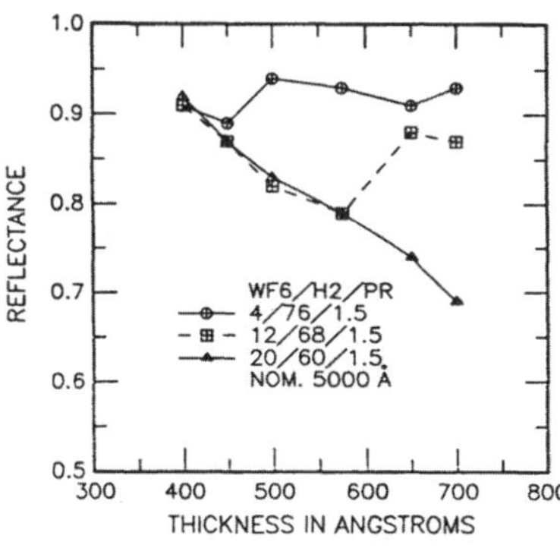

Figure 5.3 Variation of CVD blanket tungsten film reflectance as a function of film deposition temperature.[5]

Figure 5.4 Comparison of morphology of tungsten films of two thicknesses using SEM and TEM plan view: (*a*) SEM, 2200 Å; (*b*) SEM, 7900 Å; (*c*) TEM, 2200 Å; and (*d*) TEM, 7900 Å.

no growth occurs on the surrounding insulator surface. The application of this type of process is for filling very small, very deep contact holes which connect the metal layer to the substrate or one metal layer to another, without requiring any additional alignment or etching steps. The process is not sensitive to contact diameter differences but very sensitive to contact depth differences.

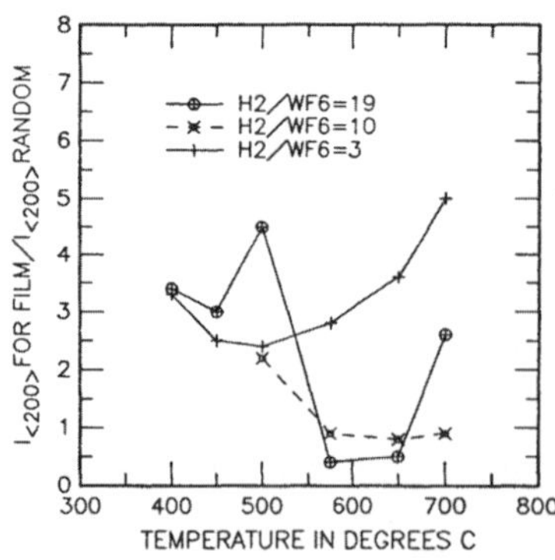

Figure 5.5 Relative <200> orientation of blanket CVD W film as a function of film deposition temperature.[5]

Selectivity Breakdown

A major problem with CVD selective tungsten is breakdown of the selectivity. It is observed that near contact regions where the deposition is taking place some deposition of tungsten occurs on the oxide surface in the form of nuclei of crystalline tungsten. This spurious nucleation may be induced by the interaction of reaction byproducts with the surface, making them activated for reaction with WF_6 to form tungsten nuclei. These conducting nuclei can cause shorts between adjacent conductors, since they are on top of the interlevel dielectric layer. The lateral size of the nuclei can be up to twice the tungsten plug depth, if the random nucleation on the oxide surface occurs early in the tungsten deposition process in the contacts.

In order for the extent of the loss of selectivity to be reduced, the insulator surface must be very clean and chemically stable. The insulator type and preparation is known to influence the stability of the selectivity of the tungsten deposition. Analysis of dielectric surfaces has been done using X-ray photoelectron spectroscopy (XPS). In most cases, UHV is required for surface analysis, so it is done ex situ to the deposition. Early work on the hydrogen reduced WF_6 process indicated that phosphosilicate glass (PSG), and glass which is plasma-deposited from tetraethoxysilane (TEOS) are good choices for improving selectivity, and that thermal oxide and silicon nitride exhibit poor resistance to loss of selectivity (LOS) of tungsten deposition.[6] In this work, the hydrogen reduction of WF_6 was used.

More recently, XPS has been used to compare the binding energy peaks of oxygen and silicon at the surface of several insulators used in microelectronics, including PSG, plasma-enhanced TEOS (PETEOS) and thermally grown oxide. In this work,[7] XPS was performed on the O $1s$ and Si $2p$ peaks using Mg K-α radiation. These peaks were found to be broader for the PSG and the PETEOS than for the thermal oxide, which can be seen in Figure 5.6. The implication is that hydroxylation plays a role in preventing loss of selectivity.

In a separate experiment, a thermally oxidized silicon substrate was treated in phosphoric acid (H_3PO_4), and XPS analyzed. The oxygen peak width was broadened toward the high-energy side, whereas the Si $2p$ peak was not. Phosphoric acid treatment is known to improve resistance to selectivity breakdown by insulators.

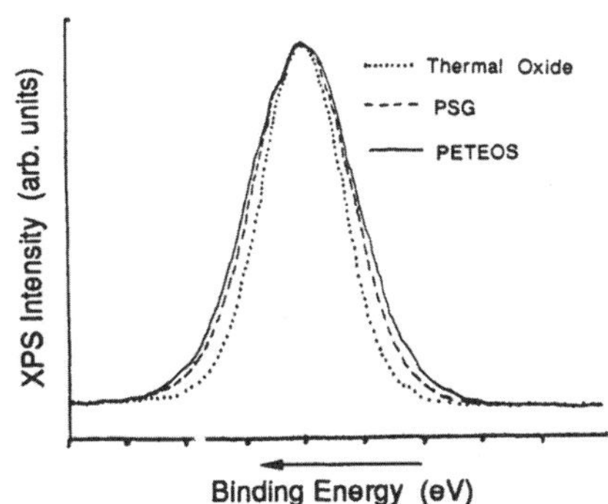

Figure 5.6   XPS binding energy spectrum of thermally grown SiO_2, PSG, and plasma enhanced oxide deposited from TEOS.[7]

The work suggests that the oxygen bonding is more important than the silicon bonding in determining resistance to selectivity loss. The less ordered surfaces are the more resistant. Work is ongoing to propose and test detailed mechanisms to explain this correlation and use the results of the studies to devise improved oxide preparation for CVD selective tungsten deposition.

Another approach to improving resistance to loss of selectivity is to obtain a fundamental understanding of the breakdown mechanism by studying the deposition process and looking for precursors to LOS. An example of analysis of selectivity breakdown is given in the work of Creighton and Rogers.[8] In their experiments, a tungsten foil was heated to 600 °C in vacuum in a WF_6 flow which impinged downstream on an oxidized silicon sample. The sample was then transferred in vacuo to a UHV system for measurement of the W $4d$ and F $1s$ XPS spectra, where it was analyzed, then heated, then analyzed again. Results indicated that a WF_x coating ($x = 3.3$) had been deposited on the room temperature substrate from the WF_6 flowing over the hot foil. The stoichiometry was inferred from calibrations using pure tungsten and pure WF_6 binding energy peaks at 243.3 and 252.5 eV, respectively, and interpolating. When the oxidized silicon sample was heated in the UHV chamber and reanalyzed, pure tungsten was observed. See Figure 5.7. These results were explained as transport of WF_5 from the hydrogen reduction reaction at the foil to the oxide surface. Here a disproportionation reaction reduced the WF_5 to less volatile WF_4. Then heating the sample further reduced the WF_4 to tungsten, by another disproportionation reaction $3WF_4 \longrightarrow 2WF_6 + W$. Once tungsten is present at the oxide surface, deposition occurs by the same hydrogen reduction process which occurs on the metal or silicon in the contacts.

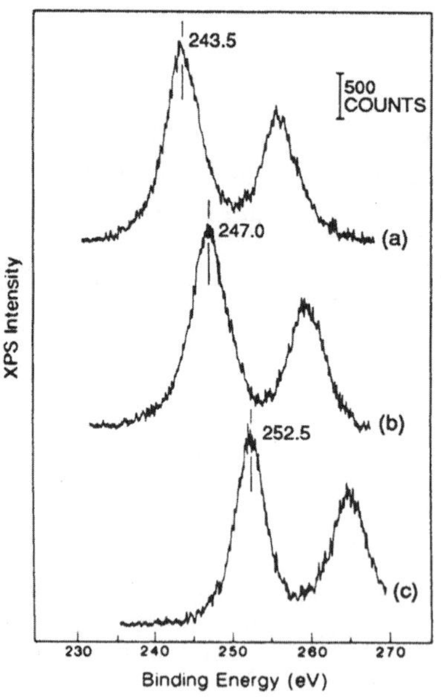

Figure 5.7   W $4d$ XPS spectra for (a) disproportionated tungsten subfluoride, (b) multilayers of tungsten subfluoride on SiO_2/Si sample, and (c) WF_6 on Ta.[8]

Current practice in selective tungsten deposition is to use the silane reduction of WF_6. This is because of its high deposition rate at low temperature (500 nm/min at 300 °C, versus 10 nm/min for hydrogen reduction at 300 °C) and its sharply reduced interaction with material at the bottom of the contact. The reduced substrate interaction is caused by the high reaction rate of the WF_6 with the silane, which supplies the WF_6 with silicon at a higher rate than the substrate can. However, this also means that making good electrical contact to the silicon can be a problem.

Substrate Interaction

The amount of substrate interaction must be minimized while insuring film adhesion and electrical conductivity of the contact. For example, silicon exposed at the bottom of a contact will reduce the tungsten hexafluoride by the reaction $2WF_6 + 3Si \rightarrow 2W + 3SiF_4$, which etches the contact surface. If this process continues locally in a contact, the metal will short through any shallow junction devices. Fortunately, this reaction is self-limiting on a clean silicon surface, and only about 200 Å of tungsten is grown, consuming about 400 Å of silicon. On the other hand, if the deposition is done prohibiting any consumption of the substrate by altering the chemistry (substituting SiH_4 for the H_2 as a reducing gas), the tungsten layer might not adhere to the silicon surface, or the resistance between the tungsten and silicon may be unacceptably high.

Another subtler effect known as "wormholing" has also been noted, in which silicon is consumed from thin but long tunnels under the contact, which are particularly noticeable in the silicon under the contact wall where the stress is at a maximum, as in the cross-sectional TEM (XTEM) of Figure 5.8.[9] This effect has been associated with the hydrogen reduction reaction, not the silicon reduction reaction which causes bulk silicon etching from the bottom of the contact. The tunnels can be many microns long and present a potentially serious diode leakage problem. These problems constitute a large part of the reason that the hydrogen reduction reaction is not currently used for selective tungsten contact filling when silicon is at the bottom.

The wormholing effect is only observable by very careful XTEM preparation. Ion milling is used to thin the samples so that electrons can be transmitted through

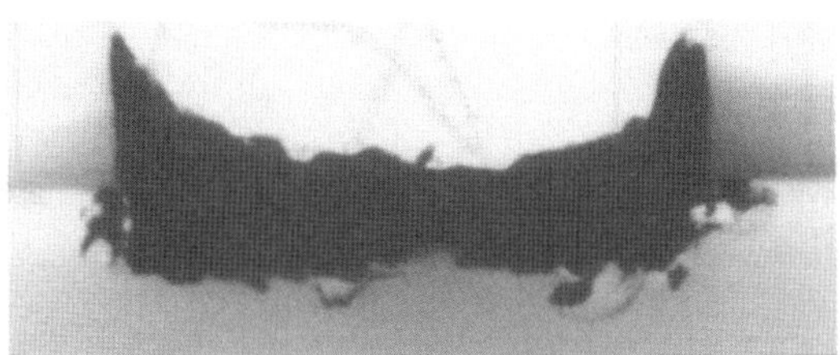

Figure 5.8   XTEM micrograph showing wormholing.[9]

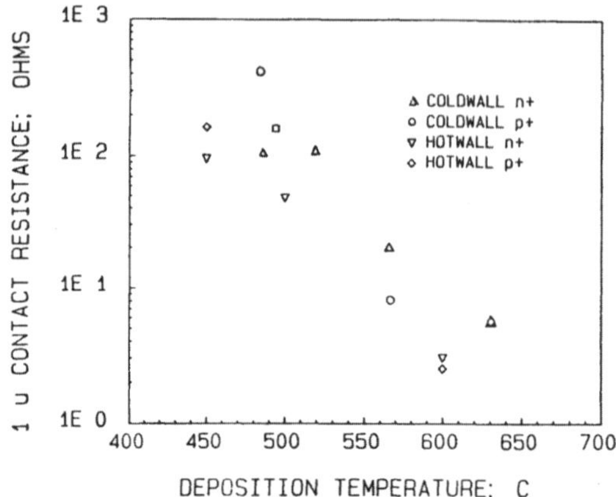

Figure 5.9   Measurement of single-contact resistance as a function of selective tungsten deposition temperature.[10]

them for imaging. One of the problems with preparation of XTEMs of tungsten as deposited on silicon is the large difference in the ion milling rates between tungsten and the silicon-based materials present in an IC. This problem results in difficulties making tungsten with sufficient transmission properties to highlight crystal structure or defects. Tungsten usually shows up "black" on a positive TEM. The wormholes, of course, are in the silicon below the tungsten contact plug. Other problems exist when the contact is to a metal or a silicide such as $TiSi_2$; WF_6 is quite reactive with metals and can form fluorides which are nonconductive and nonvolatile. These can cause loss of electrical contact. For example, aluminum trifluoride can form if selective tungsten is deposited in a via which terminates on an aluminum metal lead, preventing contact with the next level of metal.

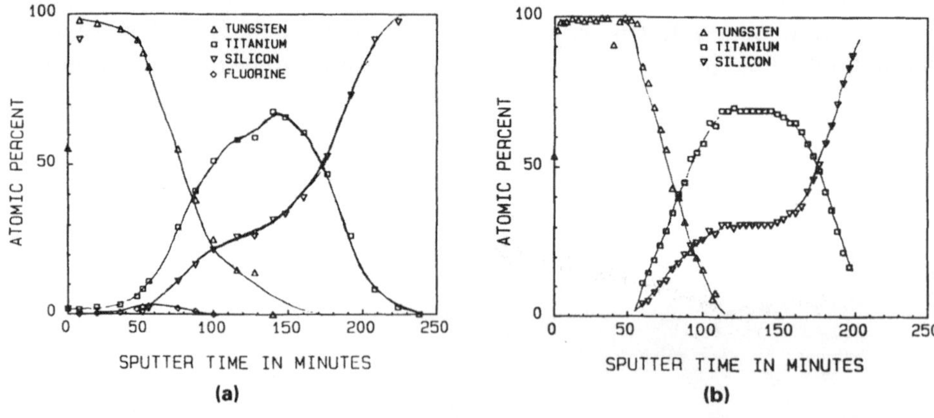

Figure 5.10   Auger depth profiles of (a) 300 °C and (b) 600 °C selective tungsten deposited on titanium silicide.[10]

   TUNGSTEN-BASED CONDUCTORS   Chapter 5

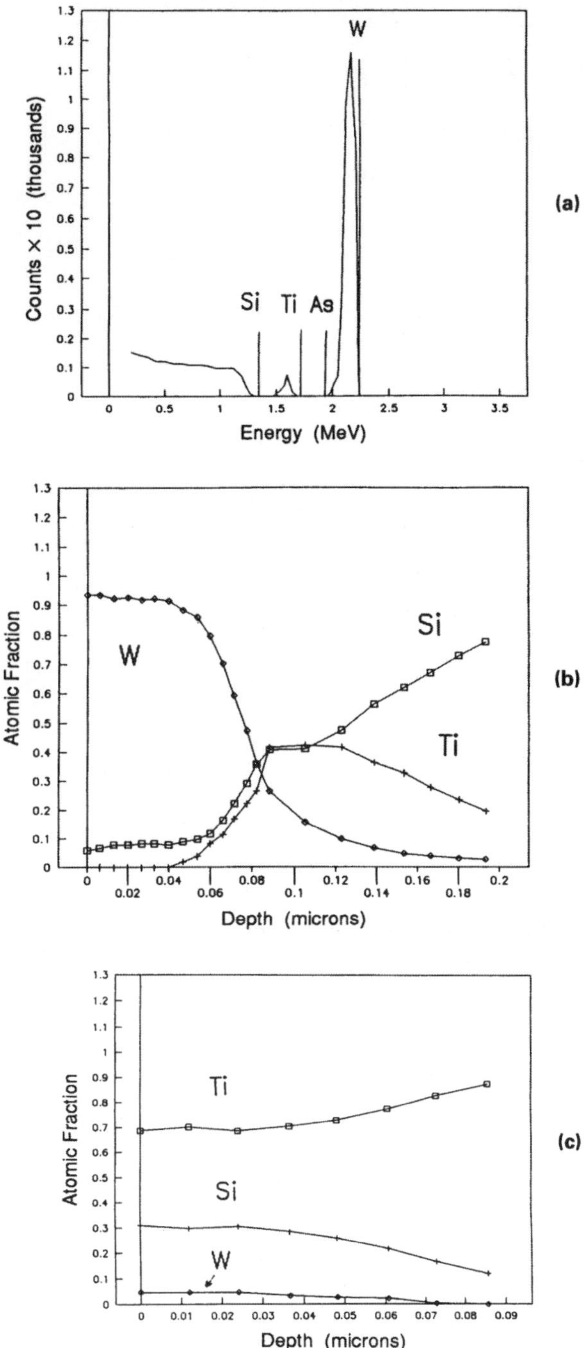

Figure 5.11 RBS (*a*) spectrum, (*b*) depth profile, and (*c*) depth profile after tungsten strip for 700-Å CVD W deposited on 1400-Å titanium silicide at 710 °C.[10]

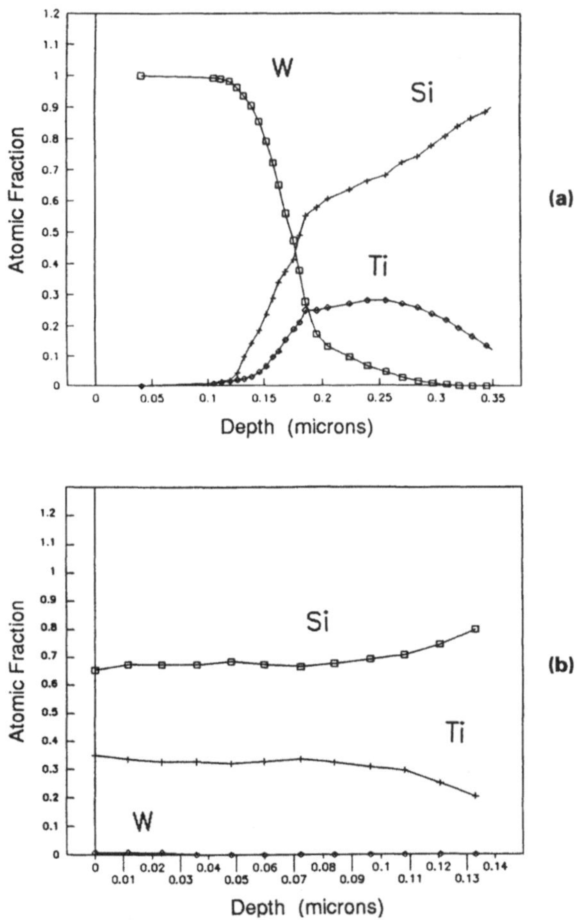

Figure 5.12   RBS depth profiles of CVD W deposited at 500 °C onto titanium silicide (*a*) before and (*b*) after tungsten strip.

TiSi$_2$ is used as a cladding on polysilicon interconnects and diffused singlecrystal silicon in order to lower its sheet resistance. CVD tungsten may be used to fill the contacts and make electrical connection. If a low-temperature hydrogen-reduced selective tungsten process is used, the formation of a nonvolatile fluoride will form at the metal–silicide interface and will lead to high-resistance contacts, as seen in Figure 5.9. In this experiment,[10] the deposition temperature of the tungsten was found to have a strong influence on the contact resistance, with temperatures above 550 °C giving acceptably low contact resistances.

The effect was studied using AES on unpatterned samples of CVD W on TiSi$_2$, using sputter etch profiling. The samples were deposited at 300 and 600 °C, respectively. See Figures 5.10*a* and 5.10*b*. Although the interfaces were not sharply delineated due to the surface roughness of the CVD tungsten, and the nonuniformity

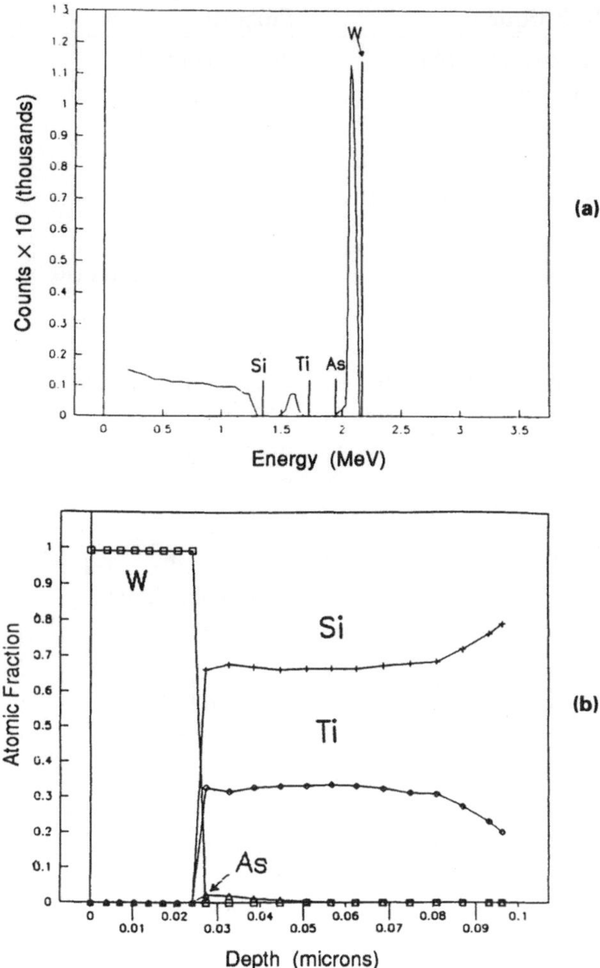

Figure 5.13 RBS (*a*) spectrum and (*b*) depth profile for a 500-Å sputtered tungsten film on 1400-Å titanium silicide annealed at 710 °C in hydrogen.[10]

of the sputter etch rate of the profiling ion beam, it is clear from Figure 5.10*a* that a fluorine peak was present at the W/TiSi$_2$ interface. Figure 5.10*b* does not show the peak, indicating that it was less than the 0.5% detection limit.

RBS was used to determine whether or not the deposited tungsten was incorporated in the TiSi$_2$. CVD tungsten and sputtered tungsten were deposited on TiSi$_2$. The CVD tungsten films were deposited at 500 and 710 °C, respectively, and the sputtered film was deposited at <200 °C and annealed at 710 °C in hydrogen. The two CVD W samples are compared in Figures 5.11 and 5.12.

The two CVD W films seem to indicate a mixed interface, whereas the sputtered film (Figure 5.13) shows a sharp interface between the W and TiSi$_2$. However, when

Parameter	Technique	Purpose
Thickness	Cross-sectional SEM Mass gain Stylus X ray Beta backscatter	Process control
Conformality	Cross-sectional SEM	Reliability
Sheet resistivity	Four-point probe Eddy current sensor	Interconnect delays (RC)
Film stress (wafer bow)	Capacitance Laser beam deflection X ray	Adhesion, Focus of lithography tools
Surface roughness	STM Reflectivity SEM AFM	Patterning, Etching
Microstructure	XRD SEM TEM	Fundamental film properties
Interfaces between films	Auger RBS SEM TEM	Junction leakage, Contact resistance
Selectivity breakdown	Optical microscopy XPS Auger TPD	Intralevel shorts

Table 5.1 Summary of concerns and methods of analysis.

the tungsten was stripped selectively from the silicide using hydrogen peroxide solution, the CVD W sample deposited at 500 °C showed no tungsten incorporation in the silicide. Here it is assumed that tungsten diffused or reacted into the $TiSi_2$ layer would not be removable in H_2O_2, as is the case with $TiSi_2$. The 710 °C deposited tungsten shows a strong tungsten signal in the silicide after peroxide etch, indicating that the deposition process incorporated it. The chemical ambient, rather than simple diffusion from the deposited layer must have been responsible for the incorporation, since the sample with annealed-sputtered tungsten on $TiSi_2$ did not exhibit mixing. The explanation for the apparently "mixed" interface in Figure 5.12*a* is the surface roughness of the tungsten film. These data suggest a deposition temperature window for the use of hydrogen-reduced selective tungsten over titanium silicide. However, other concerns, including local effects near certain types of contacts, prevent this technology from being used. Current practice is to use the SiH_4 reduction reaction for WF_6 when depositing selective tungsten.

Table 5.1 summarizes the film/deposition parameters that are critical for using both blanket and selective CVD tungsten and the analytical techniques used to characterize them.

References

1   R. Blumenthal and G. C. Smith. *Tungsten and Other Refractory Metals for VLSI Applications III*. (V. Wells, Ed.) Mat. Res. Soc., Pittsburgh, 1988, p. 47.

2   W. T. Runyan. Semiconductor Measurements and Instrumentation. McGraw-Hill, New York, 1975, p. 71.

3   G. M. Gutierrez, R. S. Blewer, and M. E. Tracy. *Tungsten and Other Refractory Metals for VLSI Applications III*. (V. Wells, Ed.) Mat. Res. Soc., Pittsburgh, 1988, p. 271.

4   S. Sivaram, S. Chen, D. Liao, R. Shukla, and D. Fraser. *Tungsten and Other Refractory Metals for ULSI Applications 1990*. (G. Smith and R. Blumenthal, Eds.) Mat. Res. Soc., Pittsburgh, 1991, p. 407.

5   R. Blumenthal, G. Smith, H. Y. Liu, and H. L. Tsai. *Tungsten and Other Refractory Metals for VLSI Applications IV*. (R. S. Blewer and C. M. McConica, Eds.) Mat. Res. Soc., Pittsburgh, 1989, p. 65.

6   D. R. Bradbury and T. I. Kamins. *J. Electrochem. Soc.* **133**, 1215, 1986.

7   R. W. Cheek, J. A. Kelber, J. Fleming, R. D. Lujan, and R. S. Blewer. *Tungsten and Other Refractory Metals for ULSI Applications 1990*. (G. Smith and R. Blumenthal, Eds.) Mat. Res. Soc., Pittsburgh, 1991, p. 99.

8   J. R. Creighton and J. W. Rogers. *Tungsten and Other Refractory Metals for VLSI Applications III*. (V. Wells, Ed.) Mat. Res. Soc., Pittsburgh, 1988, p. 63.

9   S. Tseng, L. Lane, R. Foster, S. Felch, P. Geraghty, and W. L. Smith. *Tungsten and Other Refractory Metals for VLSI Applications III*. (V. Wells, Ed.) Mat. Res. Soc., Pittsburgh, 1988, p. 303.

10 G. C. Smith, T. D. Bonifield, R. Blumenthal, J. Keenan, and P.-H. Chang. *Proceedings*. 1987 VLSI Multilevel Interconnect Conf., Santa Clara, CA, 1987, p. 155.

Barrier Films

M. LAWRENCE A. DASS

Contents

6.1 Introduction
6.2 Characteristics of Barrier Films
6.3 Types of Barrier Films
6.4 Processing Barrier Films
6.5 Examples of Barrier Films
6.6 Summary

6.1 Introduction

Very large scale integrated (VLSI) devices are made up of stacks of complex thin-film materials, which need to be chemically inert and thermodynamically stable both during device processing and throughout device lifetime. In general, the choice of materials is based on electrical properties for optimum device functioning and less on their chemical compatibility. This results in heterojunctions with large concentration gradients which are not thermodynamically stable. Thin-film materials across such interfaces either interdiffuse or react with each other in an attempt to arrive at a stable state which often is detrimental to device operation. The unwanted interdiffusion or reaction is minimized by using an additional film, called a barrier film, between two films. Barrier films are commonly used in VLSI devices in contact metallization between the substrate silicon and the aluminum interconnect metallization. They have been a key development in VLSI device processing without which the revolution in VLSI technology using Al interconnect metallization would not have been possible.[1] Both elemental and compound films are used as barriers.

The properties of barrier films depend upon their chemistry and stoichiometry, structure, microstructure, and texture, all of which can be controlled by processing conditions. It is necessary to establish a relation between processing conditions and barrier properties to arrive at a barrier with stable properties in VLSI devices. This is accomplished by characterizing the film at various stages of processing during process development. Evaluation of the barrier properties of a film is carried out by depositing the barrier and the overlayer metal films on unpatterned and patterned silicon wafers and heat-treating it to simulate the process conditions, followed by analytical examination. A variety of analytical tools are used to evaluate the chemical composition, microstructure, and chemical stability of such barriers. The use of different analytical techniques is illustrated below by referring to the most commonly used barrier films, their stability, and commonly encountered problems.

The focus of this chapter is on the barrier films used in contact metallization. These techniques are applicable to the study of any barrier film, for example, barrier films used in packaging. This chapter first describes the desired characteristics of a barrier, the kinds of barrier films, and commonly used processing methods. Then the physical, microstructural, and barrier properties of commonly used barrier films are presented.

6.2 Characteristics of Barrier Films

The function of a barrier film is to prevent any unwanted chemical reaction or interdiffusion of elements between two materials in contact, for example, a silicon substrate and aluminum interconnect metallization. The characteristics of an ideal barrier film are[2–5]

- thermodynamic stability and chemical inertness between the two materials on each side of the barrier, which were originally in contact and had the potential for reaction or interdiffusion. This inertness is required over the temperatures and time of VLSI processing and throughout the device lifetime.

- high activation energy for lattice and grain boundary diffusion.

- excellent adhesion to all the materials with which it is in contact, for example, silicon and dielectric films.

- low electrical contact resistance with the two materials between which it is interposed and low electrical resistivity.

- homogeneous in composition and free of defects.

In reality it is hard to find a thin-film material meeting all these requirements because the requirements are often conflicting. For example, a chemically inert layer does not usually form a low-resistance contact with an adjacent layer, since this often requires chemical reaction. Materials with high melting point are thermodynamically stable and chemically inert at VLSI device processing temperatures, which are typically up to 450 °C. This heat treatment reduces the contact resistance by dissolving the native oxide at the metal/substrate Si interface. Also, it reduces the metal defects and minimizes interface state densities in MOS devices. Thus, refractory metals are suitable for barrier application because of their high melting point, high activation energy for grain boundary diffusion, and low self-diffusion coefficient at room temperature. However, refractory metal films exhibit a fine-gained microstructure because of the low mobility of atoms at deposition and processing temperatures, resulting in a large grain boundary area. The activation energy for grain boundary diffusion is lower than that for lattice diffusion, and the grain boundary areas are active paths for interdiffusion across the barrier. Further, the chemical inertness results in a high contact resistance with silicon and poor adhesion with dielectric oxides. However, grain boundary diffusion and poor adhesion can be overcome by treating the barrier film prior to aluminum metal deposition and by adding a further, adhesive layer prior to barrier film deposition. The requirements for a refractory film to be stable during VLSI processing vary with film kind and microstructure, which are discussed in the sections on different barrier materials.

6.3 Types of Barrier Films

A barrier can be formed between a substrate and metal, which would otherwise interact, by using either a film that reacts with its adjacent layers, thus providing an alternate reaction path, or a film that completely prevents any interdiffusion. These are called sacrificial and diffusion barriers,[6] respectively. Barrier films used in VLSI processing come under both categories. For example, Ti, Mo, V, W, and Cr are sacrificial barriers, since they react with both substrate and interconnect metal, whereas TiN, W–Ti(N), ZrN, and RuO_2 are diffusion barriers, since they do not react with adjacent layers. The stability of a sacrificial barrier depends upon the remaining presence of an unreacted part of the barrier film, either on the substrate or the interconnect metal side, through which Si diffusion is very small. However, the reaction of a sacrificial barrier film with aluminum could cause metal voiding, which is a reliability problem. For example, Cr is less attractive as a barrier film because it results in metal voiding although it has good barrier properties. Nitrogen out-diffusion from compound barrier films—such as TiN—into the metal during the subsequent VLSI processing is a potential reliability issue. Hence, a barrier film is chosen based on its barrier properties *and* its compatibility with the interconnect metal *and* on the rest of the processing steps and wafer throughput.

6.4 Processing Barrier Films

Barrier films are commonly processed by sputtering in an inert or reactive atmosphere, depending upon the target and type of film. These are called inert and reactive sputter deposition. During deposition, the substrate may be held at higher temperature or under a bias to alter the film properties. This is commonly used to process films of compound materials. Barrier films are also processed using chemical vapor

                                                        BARRIER FILMS   Chapter 6

deposition (CVD) to improve the step coverage at contacts with high aspect ratios. This is becoming more important with the drive to decrease the feature size and to increase the density of devices on the chip.

Inert Sputtering

An elemental or a compound target is sputtered in an inert gas atmosphere to deposit elemental or compound films on the substrate. During this process, the sputtered species do not react with the gas in the chamber. The composition of the sputtered film may be different from the target composition, due to differential sputtering and resputtering of the elements present. For example, the composition of a W–Ti film is deficient in Ti compared to the target composition.[7] Compound films are also processed by cosputtering elemental targets in an inert atmosphere, for example, W and Ti targets to form W–Ti. Compound target sputtered films tend to have higher stress due to poor mobility of the sputtered species on the substrate. Higher stresses may lead to loss of adhesion between the films, resulting in substrate and film cracking.

Reactive Sputtering

Reactive sputtering is a versatile compound thin-film deposition technique in which an ultrapure elemental target is sputtered in a reactive gas/argon atmosphere. The composition of the film can be changed by controlling the partial pressure of the reactive gas in the sputtering chamber.[8, 9] During reactive sputtering, reactive gas flow and its partial pressure exhibit a nonlinear relationship. Figure 6.1 shows this relationship for the reactive sputtering of Ti in a nitrogen ambient to form TiN. At the initial stages of sputtering, partial pressure does not increase with the increase

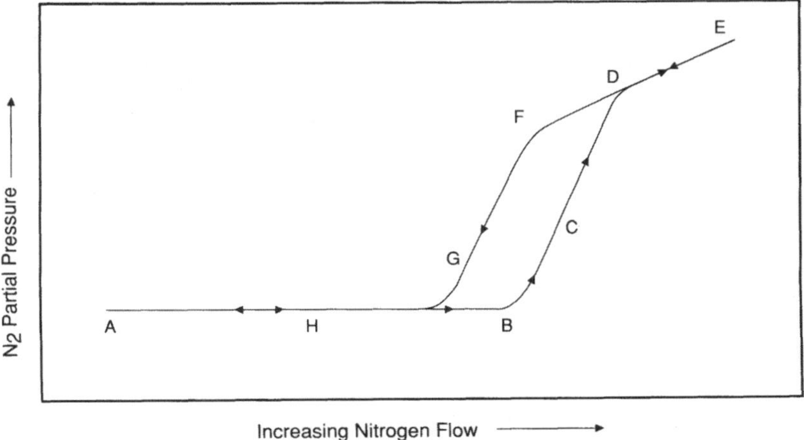

Figure 6.1 **A schematic of the relationship between nitrogen partial pressure and nitrogen flow illustrating the hysteresis observed during the reactive sputtering of TiN.**

in gas but remains nearly constant along line AB because of gettering of nitrogen by sputtered Ti; a substoichiometric film is deposited under this condition. A further increase in gas flow results in a rapid increase in gas partial pressure along line BCD because of the reduction in N_2 consumption caused by compound formation at the target surface, which results in a decrease in Ti sputter yield. This is known as "target poisoning"; it is accompanied by a decrease in sputtering rate. As the gas flow continues to increase, partial pressure increases along line DE, where super-stoichiometric TiN is deposited. When the gas flow decreases, the relationship between gas flow and partial pressure traverses DFGH to the initial state, resulting in a hysteresis. The shape of the hysteresis loop can be altered by changing sputtering conditions and the geometry of the sputtering chamber. Decrease in gas flow is an effective way to "deposition" a target. Stoichiometric TiN can be processed at point B and also at point D in the hysteresis loop, but at a much lower sputtering rate because of compound formation on the target surface at D.

A high sputtering rate is desirable to increase the throughput of wafers processed. When the target power is changed to increase the sputtering rate, a different partial pressure of reactive gas is required. In a sputtering chamber there are apparent pressure differences because of gas heating and rarefaction in front of the target. As the power is applied to the target, the density of the gas in front of the target is reduced because of gas heating from collisions with energetic ions and neutrals and rarefaction due to metal atom transport. However, the overall effect is constant pressure because the metal atoms flowing to the substrate are in a gaseous state and make up a significant portion of the local pressure at higher rates. The composition of the reactive sputtered film is affected because of the reduction in arrival rate of the reactive species at the substrate due to the apparent pressure or density change in front of the target. In order to maintain stoichiometry, the overall partial pressure of the reactive gas must be increased to compensate for the decrease in density of reactive gas pressure in front of the target as the power is increased to compensate for the increase in density of the metal atoms. Some examples of reactively sputtered films are TiN, WN, W–Ti–N, and ZrN.[1–10]

Chemical Vapor Deposition

The step coverage of most sputter-deposited barrier films is poor due to the low mobility of sputtered species, especially at the contact bottoms, and weak spots are observed at contact corners, for example, in reactively sputtered TiN. Alternate film deposition techniques, such as CVD, are actively pursued to improve film step coverage and to process a film with uniform composition, at contact geometries having a high aspect ratio. Two different precursors are used for CVD deposition of TiN. A Ti-containing compound, such as $TiCl_4$, is reduced by a mixture containing H_2–N_2 or NH_3–Ar in a low-pressure system.[11] Alternately, an organometallic precursor, such as tetrakis(dimethylamino)Ti, is reduced with an NH_3 and Ar mixture to deposit MOCVD TiN with a much better step coverage than PVD films.[12] The presence of contaminants such as Cl in $TiCl_4$-based chemistry films and O and C in organometallic precursor-based TiN films degrade the quality of the film. Further, these reactions take place at high temperatures, for example, above 600 °C for $TiCl_4$ chemistry and 450 °C for organometallic-precursor chemistry, which may be an issue with process integration. CVD techniques are also prone to produce particles during deposition.

Nitridation and Rapid Thermal Annealing

Thin TiN films can be processed at the surface of Ti by rapid thermal annealing (RTA) a Ti film at 900 °C for 30 s in a nitrogen ambient. During the annealing, N_2 reacts with Ti to form TiN at the surface, and the TiN thickness is limited by the diffusion of N_2 through TiN to TiN/Ti interface. This is a self-limiting reaction because of the high activation energy for diffusion through TiN. At contacts, where Ti is in directly on Si, both react to form a silicide during annealing. This metal scheme has a potential for making structures having silicided contacts, with built-in diffusion barriers and shallow junctions. Substoichiometric nitride and oxides form during RTA, depending upon annealing ambient and temperature.[13] The barrier properties of the resultant TiN would be limited by the thickness and homogeneity of the TiN formed by RTA.

6.5 Examples of Barrier Films

In the early stages of diffusion barrier film development, elemental films such as Ti were used. These were replaced with compound films because of the detrimental reaction between Ti and its adjacent layers. Although compound films are stable during device processing, the absence of reaction between the substrate Si and the barrier resulted in high contact resistance, and the barrier's adhesion with Si and glass substrates was poor. However, an optimum barrier is achieved by combining both elemental and compound barrier films to obtain a stable barrier film with low contact resistance and good adhesion with Si and glass substrates, such as Ti/TiN. Electrical testing of a fully or partially processed device with a barrier film integrated metal structure is a final test to establish the barrier film stability. During development, when a barrier film fails to meet electrical performance criteria, one cannot identify the causes of the failure using electrical tests. This requires a knowledge of physical properties and the relationship between microstructure and processing parameters, as well as the study of thin-film reactions between the barrier film and adjacent layers.

Thin-film reactions can be studied using many different analytical techniques. Common analytical techniques such as Rutherford backscattering spectrometry (RBS), Auger electron spectroscopy (AES), secondary ion mass spectrometry (SIMS), and cross-sectional transmission electron microscopy (XTEM) are presented below in the study of barrier film reactions and stability. Physical properties, microstructure, and barrier properties of three commonly used barrier films, Ti, W–Ti–(N), and TiN, are analyzed. Some salient characteristics of these films are

Film	Barrier Type	Barrier Thickness, nm	Advantages	Disadvantages
Ti	Sacrificial	50 nm	Ohmic contact layer; stable up to 400 °C/30 min	Reacts with adjacent layer and can fail
TiN	Inert	70 nm	Stable up to 500 °C/30 min	Poor step coverage
W–Ti(N)	Inert	100 nm	Excellent step coverage; stable up to 500 °C/30 min	Particles; mobile ion contamination from target

Table 6.1   Some properties of commonly used barrier films.

presented in Table 6.1.[14] The thin-film reaction of Ti films is studied by using RBS and XTEM, that of W–Ti films by AES, and that of TiN films by SIMS and XTEM. The choice of any particular analytical technique is not limited by the film kind but depends upon the nature of the study. The reader is referred to the Appendix in this volume for a brief description of these techniques or, for a full but nonspecialist description, to the lead volume of this series, *Encyclopedia of Materials Characterization*.

Titanium Thin Films

Physical properties   Titanium thin films are deposited by dc magnetron sputtering a pure Ti target in an argon atmosphere. As-deposited Ti films appear metallic gray. Ti is a refractory metal with a density of 4.5 g/cm^3 and a melting point[15] of 1663 °C. The crystal structure is a hexagonal close-packed structure with lattice parameters a = 2.95 Å and c = 4.686 Å. The thin-film resistivity of a Ti film sputtered in a clean environment is ~50 μΩ·cm.

It is essential to measure and to monitor the thickness of the film accurately because barrier properties depend upon the film thickness. The thickness of the film can be measured using sheet resistance, RBS, X-ray fluorescence (XRF), cross-sectional imaging, and step height measurement or profilometry. The accuracy of the measurement by both sheet resistance and RBS depends upon the bulk material properties. The ratio of bulk resistivity of pure titanium to sheet resistance is an approximate measure of film thickness. In the case of RBS, the ratio of areal density (derived from RBS spectra) to bulk density of Ti is a measure of the thickness. Sheet resistance and RBS thickness values may be in error since they do not take into account the physical film properties. Sheet resistance measurements assume bulk resistivity and physical film characteristics—grain size, defect concentration, film density, impurity content, etc.—are constant across the wafer, whereas RBS thickness measurements assume a film density. In some situations, the physical film characteristics can vary substantially across the wafer but still give rise to the same nominal sheet resistance or nominal RBS thickness. Hence, an accurate measurement of the *physical* thickness is necessary to calibrate film thicknesses determined by sheet resistance or RBS for particular processing conditions.

XRF is another method for measuring mass thickness rather than physical thickness. However, a control chart, in which the ratio of the Ti peak to the Si peak from the substrate for different thicknesses of Ti film is plotted against the accurate film thickness measured by some absolute method, can be used to measure the thickness of an unknown film. This is often used as a control monitor in a manufacturing environment. XTEM and scanning electron microscopy (SEM) can be used as absolute methods to measure the physical thickness of film, XTEM being the more accurate of the two because of its superior resolution.

A more direct technique, stylus profilometry scanning over a step, can provide a physical film thickness. The accuracy of this technique for thin films is poor. However, the atomic force microscopy (AFM) technique can be used to measure the thickness of a thin film at a step with much higher accuracy (10–20 Å). AFM output is presented as a two-dimensional scan, in contrast to the one-dimensional scan obtained from stylus profilometry. Hence, an average film thickness can be determined by measuring the step height at locations within an AFM scan area. AFM can be operated in ambient conditions and does not need any special specimen preparation. Atomic force microscopes, which can handle a whole wafer for use as in-line monitors, are commercially available. Since most of the AFMs are currently operated by a fine-pointed tip in contact with the sample, possible damage of the sample and convolution of the tip geometry in the image are of concern. Profilometry requires step-etching, which can be carried out by depositing Ti film on oxide; this can sometimes be a problem.

Microstructure Titanium film growth takes place by nucleation and grain growth. Microstructure of the film is affected by the process conditions, such as target power, chamber pressure, substrate temperature, and film thickness. For example, a 1000-Å-thick Ti layer under aluminum metal deposited on a nonheated substrate exhibits equiaxed grain morphology, as shown in Figure 6.2*a*. In contrast, a 600-Å-thick Ti film deposited on a heated substrate exhibits larger grain size compared to the room temperature film and has columnar grain boundaries, where

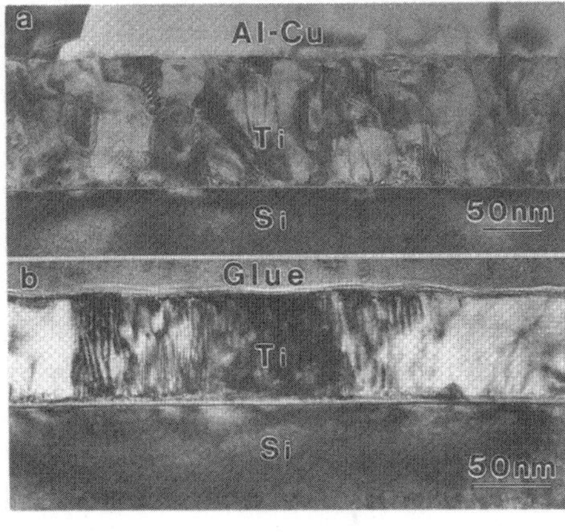

Figure 6.2 XTEM images of sputtered Ti films deposited on Si substrates held at (*a*) room temperature and (*b*) 300 °C.

grain boundaries extend from the substrate to the surface, as shown in Figure 6.2*b*. Higher substrate temperature is accompanied by an increase in mobility of the sputtered species, resulting in grain growth and columnar grain boundaries. The differences in the microstructure can influence the barrier properties—for example, columnar grain boundaries can act as rapid diffusion paths. The average grain sizes of the Ti films deposited at room temperature and on the heated substrate are about 300 and 450 Å, respectively, for the silicon shown in Figure 6.2. On exposure to atmosphere prior to aluminum deposition, a very thin native oxide layer of TiO_2 is formed on the surface, which also influences the barrier properties.

Barrier properties   A common metallization scheme using a Ti barrier film is obtained by depositing a desired thickness of Ti on substrate silicon prior to Al metal deposition. The native oxide on Si is reduced by Ti, resulting in low-resistance contacts and reducing the specific contact resistivity during subsequent heat treatment cycles. In addition, processing steps used to clean the substrate silicon affect the contact resistance and adhesion of Ti with underlayers. Ti is a sacrificial barrier because it reacts with adjacent films during postmetallization alloying typically at 400 to 450 °C for 30 min. Formation of an amorphous (Ti,Si) solid solution and C49 phase of $TiSi_2$ has been observed. Intermetallic compound formation at the Ti/Al interface is observed above 300 °C and has been studied extensively. The thin-film reaction follows parabolic growth kinetics,[17] with the amount of Ti consumed being given by

$$x = t^{1/2} d_0^{1/2} \exp(E_g/2kT) \tag{6.1}$$

where x is the thickness of Ti consumed, t is the thickness of Ti, $d_0 = 0.15$ cm^2/s, E_g, the activation energy for the reaction, is 1.85 eV, k is the Boltzmann constant, and T is the temperature in kelvin.

To calculate the thickness of Al consumed, one can use Equation 6.1 with $d_0 = 1.2$ cm^2/s, and one can calculate the thickness of Al_3Ti produced using $d_0 = 2.0$ cm^2/s. The step coverage of Ti in contacts is very poor due to the low mobility of Ti atoms at processing temperatures. One must take into account both the interaction of Ti with Al and its poor step coverage characteristics when selecting the thickness of the film. Pure Ti is a good diffusion barrier, because the diffusivities of Si and Al through Ti are very low. But in the presence of the Al_3Ti intermetallic, Al diffuses through the aluminide phase and piles up near the substrate Si, creating a Si/Al interface. This is harmless, since Al does not dissolve in Si but Si diffuses into Al due to the solid solubility of Si in Al. Transport of Si results in the formation of etch pits in the substrate Si. The etch pits fill with Al, forming spikes that are conductive and that may penetrate a junction, causing electrical failure.

The reaction kinetics of intermixing the barrier with adjacent layers can be examined using RBS, AES, and SIMS. These techniques can be used effectively to compare the extent of reaction or intermixing between the as-processed and heat-treated samples. RBS spectra of as-deposited and 450 °C, 30-min heat-treated Si/PtSi/Ti/Al samples are shown in Figure 6.3.[18] A PtSi layer is used to increase the

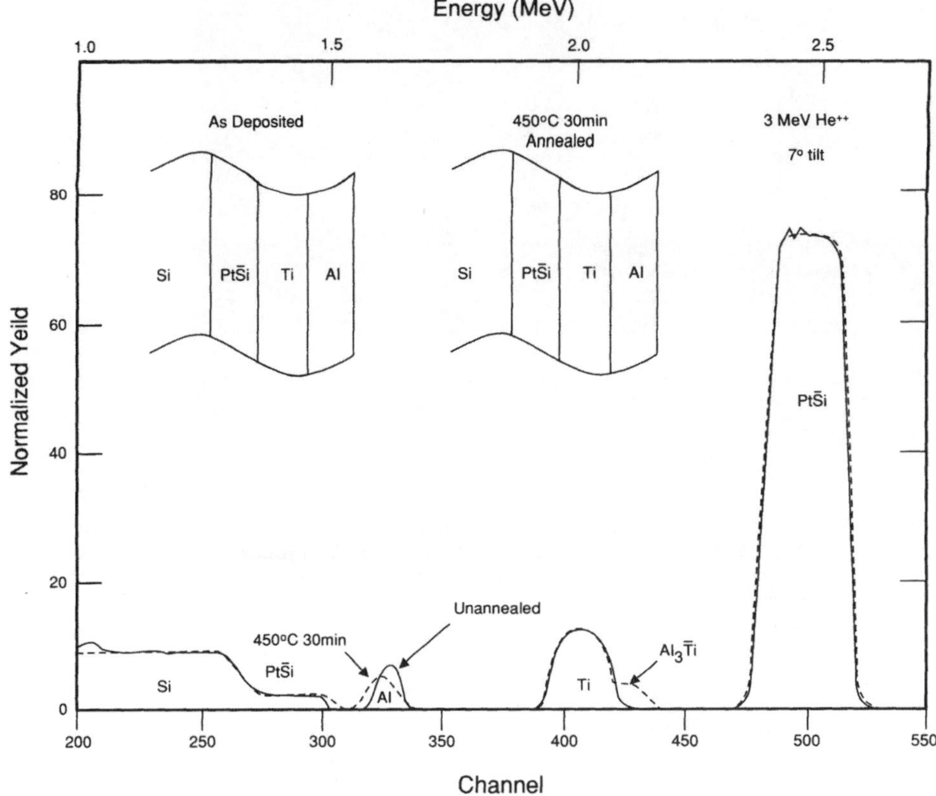

Figure 6.3 RBS spectra from unannealed (*continuous line*) and 450 °C, 30-min, annealed (*dotted line*) Si/PtSi/Ti/Al metal structure samples.

Schottky barrier height on contact with *n*-Si. When in direct contact, Al and PtSi react at 300 °C to form $PtAl_2$, which reduces the barrier height—an undesirable effect. A Ti barrier film is interposed between the PtSi and Al layers to minimize the detrimental reaction between the two. In Figure 6.3, the shoulder to the right of the Ti peak of the alloyed samples is a clear indication of the reaction between Ti and Al; the reaction compound is deduced to be Al_3Ti. The reaction of Al is also seen by the shift of the Al peak to the left, closer to the substrate, and a decrease in the peak height in the annealed sample. The absence of any shift at the left side of the Ti peak and right side of the PtSi peak between the as-deposited and heat-treated spectra indicate that Ti and PtSi do not react under this condition and that Ti is an effective barrier. RBS can be used to deduce the extent of reaction at an interface when the reaction between two films is widespread or homogeneous, for example, Ti and Al. A spatially resolved method such as XTEM must be used to confirm that the reaction is homogeneous. AES and SIMS, when used in the depth-profiling mode, will give similar information to RBS but with much better depth resolution. The drawback is that when used in this manner these techniques are

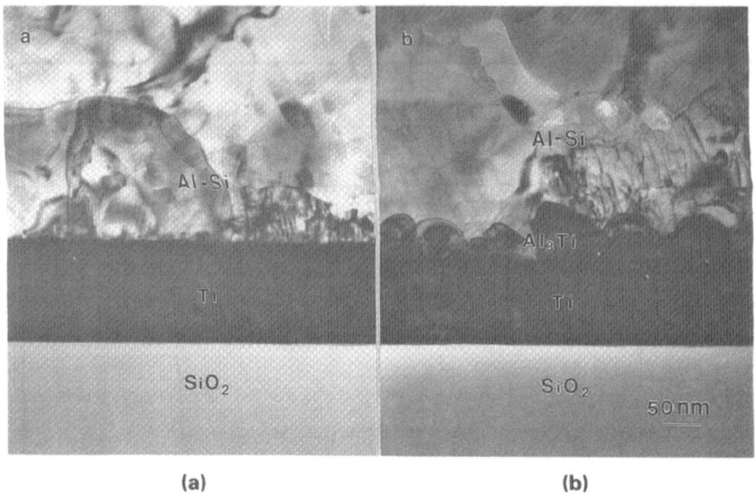

Figure 6.4   XTEM images of (*a*) as-deposited and (*b*) 460 °C, 40-min annealed Ti/Al–Si metal structures on oxide substrate. The rough Ti/Al–Si interface in the annealed sample is from Al₃Ti intermetallic compound formation.

destructive, and their data is more complex and difficult to interpret. However, they can be used in a spatially resolved manner to check for lateral inhomogeneity.

The interaction between the barrier and the aluminum can be examined microstructurally using XTEM. XTEM images of as-deposited and 430 °C, 70-min annealed SiO_2/Ti/Al–2% Si samples are shown in Figure 6.4. The Ti/Al–2% Si interface is well-defined and sharp in the as-deposited sample, and the corresponding interface in the annealed sample is rough, indicating reaction between Ti and Al–2% Si. The compound formed is intermetallic Al_3Ti. The intermetallic is finegrained and exhibits globular morphology. The grain size of Al_3Ti is smaller than the Ti grain size, thus increasing the grain boundary areas for Al in-diffusion into Ti. The metallization using only a Ti barrier would eventually fail unless alternate steps, such as two-stage Al deposition, were used.

Intermetallic compound, Al_3Ti formation at the Ti/Al interface is influenced by alloying elements present in aluminum. Silicon present in Al–1% Si was observed in Al_3Ti at interstitial positions formed during alloying at 450 °C for 30 min.[19] No intermetallic other than Al_3Ti was observed, because the solid solubility of Si in Al_3Ti may be as high as 15%. The effect of Cu in aluminum on Al_3Ti intermetallic formation has been observed to be very different from that of Si in Al. Addition of 3 wt% Cu in aluminum retards the growth rate of Al_3Ti by an order of magnitude at 400 °C. Evaluation of the kinetics of the reaction showed that the activation energy of the reaction is increased from 1.8 ± 0.1 eV for unalloyed aluminum to 2.4 ± 0.1 eV for Al–3 wt % Cu.[20]

Tungsten–Titanium Thin Films

Physical properties   Tungsten–titanium films processed by magnetron sputtering result in a metastable solid solution because of the negligible equilibrium solid solubility of Ti in body-centered cubic (bcc) tungsten at room temperature. W–Ti is either sputter-deposited from a W–Ti alloy target or cosputtered from W and Ti targets in an argon atmosphere. As often observed in thin films, the magnetron sputtering process results in supersaturation of Ti in a bcc tungsten matrix. The nominal composition of the W–Ti barrier film commonly used is 80 at. % W and 20 at. % Ti. The composition of the sputtered film is dependent upon the target–substrate distance, target power, and the argon gas pressure. By appropriate control of process parameters, the composition and sputtering rate can be controlled. The crystal structure of the W–Ti film is bcc. A number of physical properties such as lattice constant, density, stress, and resistivity are dependent upon the Ti content in W and the sputtering conditions. A bias voltage is applied to the substrate to effectively alter the film properties. A beneficial effect on the barrier properties of a W–Ti film has been observed when it is sputtered in an N_2/Ar atmosphere. Such W–Ti films exhibit high film stress which is dependent upon the composition. For example, stress in pure $W_{80}Ti_{20}$ is compressive at about 1.25 GPa, but stress is tensile at about 2.5 GPa for $W_{50}Ti_{15}N_{35}$. The nitrogen in W–Ti–N occupies an interstitial position. The sputtered W–Ti film composition in an argon atmosphere is deficient in Ti compared to the target composition, because of effective resputtering of Ti atoms from the film by energetic neutral Ar atoms.[7]

Microstructure   The structure and microstructure of W–Ti films are dependent upon the processing conditions. The lattice parameter of a W–Ti film is influenced by both sputtering atmosphere and substrate bias. The lattice constant increases on addition of Ti to W due to the expansion of the W bcc lattice from 3.1656 to 3.1818 Å for a nominal $W_{80}Ti_{20}$ film deposited on an unbiased substrate in an argon atmosphere. This is because Ti occupies interstitial sites in the W lattice. W–Ti films exhibit a strong texture along <110> parallel to the growth direction, which becomes stronger on addition of N in the sputtering chamber. When 15% nitrogen is introduced in the sputtering chamber, both reduction in lattice constant to 2.9897 Å and incorporation of N in the W–Ti film are observed. On application of a high bias to the substrate, the lattice constant of the $W_{80}Ti_{20}$ film changes from 3.1678 Å at 300 V to 3.2244 Å at 700 V because of incorporation of Ti into the W lattice.[21] At low substrate bias deposition of W–Ti films, Ti is present as second phase particles, and the lattice parameter remains similar to that of pure W. X-ray diffraction (XRD) analysis of these films can provide interplanar spacings, lattice constants, and information about the texture, but XRD is not sensitive enough for detection of a second phase when present in small amounts (1 to 2 vol %), for example, Ti particles in the W–Ti.

A plan view TEM image of $W_{80}Ti_{20}$ is shown in Figure 6.5. Electron diffraction revealed a homogeneous bcc phase. The microstructure of the W–Ti film consisted

Figure 6.5 Planar TEM image of a sputter-deposited $W_{80}Ti_{20}$ film exhibiting laminar features within grains.

of equiaxed small grains and elongated large grains. Within a grain, very closely spaced bands of wide and narrow lines are observed which are characteristic of W–Ti films. This could be due to agglomeration of grains forming one large grain with the same orientation. Cross-sectional examination of W–Ti films shows very fine grains and agglomeration of grains in which columns having nearly the same orientation are present. The microstructure of the films transforms to very finegrained morphology when W–Ti is sputtered in the presence of N. When a $W_{80}Ti_{20}$ film deposited on a unbiased substrate is annealed at 450 °C for 30 min, phase separation into equilibrium phases of W and Ti takes place. In the case of similar films deposited on biased substrate, a decrease in lattice parameter is observed on annealing at 450 °C for 1 h, as determined from XRD.

Barrier properties W–Ti films exhibit excellent adhesion with Si and glass substrates and very good step coverage. When a W–Ti film is brought in contact with substrate silicon, as a barrier between Si and below Al, the W–Ti film does not form a low-resistance contact. A thin film is therefore interposed between Si and W–Ti to minimize the specific contact resistance. For example, W–Ti deposition is preceded by deposition of a thin Ti film, which reacts with substrate Si during subsequent heat treatments resulting in ohmic contact. Another commonly used process is to preform PtSi at contacts prior to W–Ti deposition.

Chemical characteristics and the relative distribution of elements present in thin-film metal structures can be examined by depth profiling techniques, such as AES and SIMS. AES is a very surface-sensitive technique, since the Auger electrons analyzed emanate from the top 10 to 20 Å of film surface. Auger peak shapes for each element are very distinctive, and changes in element chemical state can be determined (see the article on Auger spectroscopy in the *Encyclopedia*). Information on a lower depth in the metal structure is obtained by consecutively sputtering off layers of material by argon beam. Auger signals collected at different depths are plotted against time or distance to obtain Auger depth profiles.[22] Artifacts during

BARRIER FILMS   Chapter 6

depth profiling, such as preferential sputtering and mixing effects at the interface, reduce the depth resolution. Reliable results can be obtained by depth profiling a control sample along with the heat-treated sample and comparing the results. The depth resolution during depth profiling can be improved by using a model metal structure with thinner Al metal—thinner as compared with actual design thickness—since sputtering through Al is known to result in increased roughening of the surface. Rotating the sample during sputtering (Zalar rotation) also improves depth resolution.

Plots of Al, W, Ti, Si, and O Auger signals monitored during AES depth profiling through an as-deposited stack of Al–Si/W–Ti/Ti on Si and 460 °C heat-treated samples are shown in Figure 6.6. An underlayer Ti below W–Ti is used here to obtain contacts with low specific contact resistance. The onsets of W and Si signals are advanced closer to the surface in the annealed sample compared to the as-deposited sample, indicating reactions at Al–Si/W–Ti and Ti/Si interfaces. The detailed Auger line-shapes corresponding to different locations in the depth profiles are characteristic of the compound formed and can provide information about the nature of the reaction compound. However, the Auger depth profiles have to be calibrated using the results from other studies, such as TEM and RBS. Microstructural studies (diffraction) have shown that W–Ti and Al react to form WAl_{12} and Al_3Ti at temperatures above 400 °C.[21] The barrier breakdown takes place by diffusion of Al through the intermetallic compounds and the grain boundaries of W–Ti.

Alloying aluminum with small amounts of Cu has been found to have no effect on the barrier properties of W–Ti between Si and Al.[23] Improvements in the W–Ti barrier properties is observed by contaminating the barrier film prior to metal deposition. This is carried out either in situ by sputtering in Ar/N_2 or ex situ by exposing the W–Ti film to the atmosphere prior to metal deposition. It is postulated that, when W–Ti is exposed to air, atmospheric oxygen adsorbed at the grain boundaries oxidizes the unreacted Ti particles, thus delaying the reaction between the barrier and aluminum metal during annealing. Superior barrier properties are observed with W–Ti–N films, since they are free of elemental Ti particles and voids, compared to W–Ti film exposed to atmosphere only prior to metal deposition. The barrier properties of W–Ti–N have been found to improve with increase in N_2 content. The stability of W–Ti–N films up to 475 °C has been explained by two different mechanisms, where both are supported with experimental results. In the first case, the W–Ti–N film is made of finely dispersed compounds such as β-W, TiN, and Ti_2N in a W–Ti matrix, which are stoichiometric and refractory and do not react with the metal.[21] Alternately, N in the W–Ti–N reacts with Al to form a thin impermeable continuous layer of AlN at the barrier and metal interface, which has very high activation energy for diffusion preventing any interdiffusion.[24]

Titanium Nitride

Physical properties Titanium nitride, TiN, is a hard refractory material extensively used as a wear- and corrosion-resistant coating over a period of time. Thinner

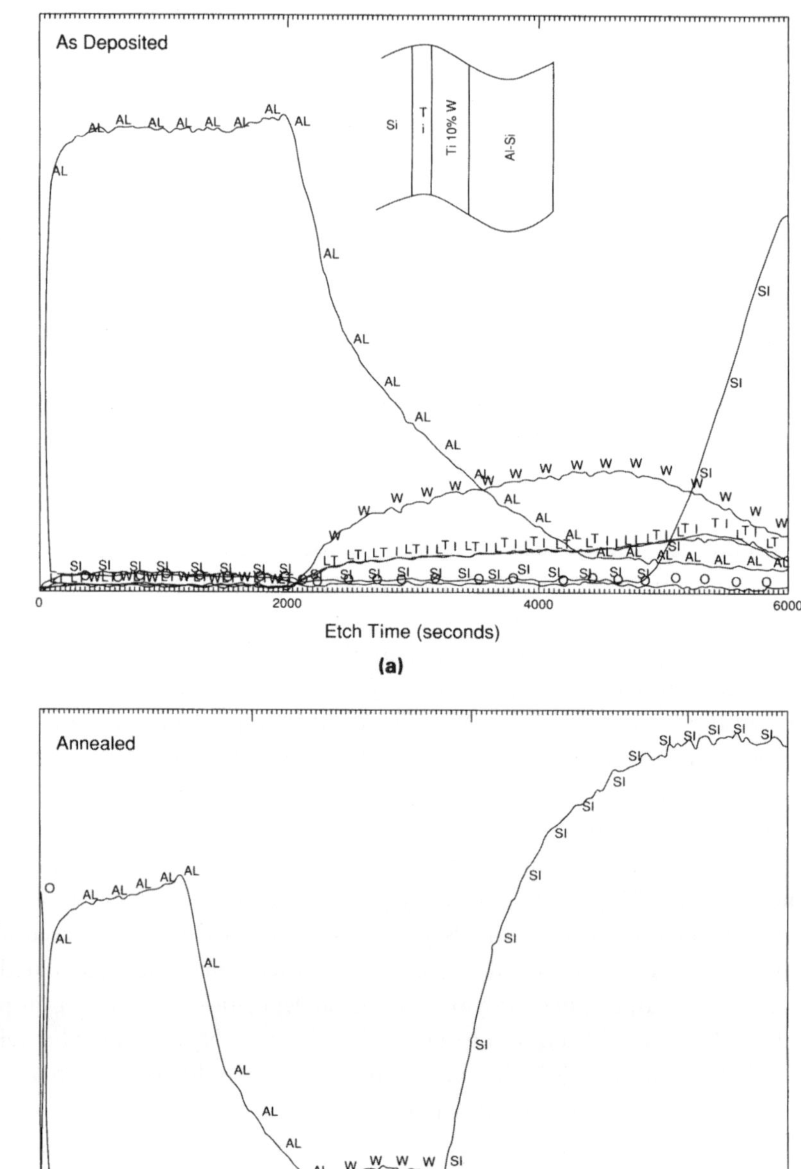

Figure 6.6   AES depth profiles through (*a*) as-deposited and (*b*) 460 °C annealed Ti/W–Ti/Al–Si metal structures on Si.

films of TiN have found use in VLSI devices as barriers and antireflective coatings. The crystal structure of TiN is face centered cubic (fcc) with a NaCl structure and a lattice constant of 4.23 Å.[25] The bulk density and melting point of TiN are 5.22 g/cm^3 and 2930 °C, respectively.[15] TiN is a defect compound owing to the presence of vacancies. The sheet resistance of a reactively sputtered TiN film is a strong function of the sputtering conditions, stoichiometry, and composition, varying from 25 to 200 μΩ·cm. The golden yellow color of TiN can be used as an indication of the stoichiometry of the film, as described below.

The characteristic color of some metals is due to preferential absorption in the visible spectrum, depending on electronic structure.[26] Hence, the color of TiN depends upon both composition (stoichiometry) and lattice volume (the volume of the unit cell), since these define the electronic structure. The lattice volume of a unit cell is affected by deposition rate, defects, trapped Ar, and alloying treatment. A typical reflection curve of TiN compared to the noble metals Ag, Au, and Cu[26] is shown in Figure 6.7. The color of TiN is golden yellow due to the depression of the reflectance at the blue end of the spectrum, as shown in Figure 6.7. Based on a simple ionic model, a reduction in the nitrogen concentration shifts the plasmon edge to higher frequencies, making the film more silver, whereas an increase in the lattice volume shifts the plasmon edge in the opposite direction, making the film more golden. Hence, the stoichiometry of the film can be inferred qualitatively by visual inspection. The thickness of thin TiN films of less than 40 nm used as antireflective coatings is measured by optical interference techniques using visible radiation. Thicker films of TiN are used as barrier films, and the film thickness is

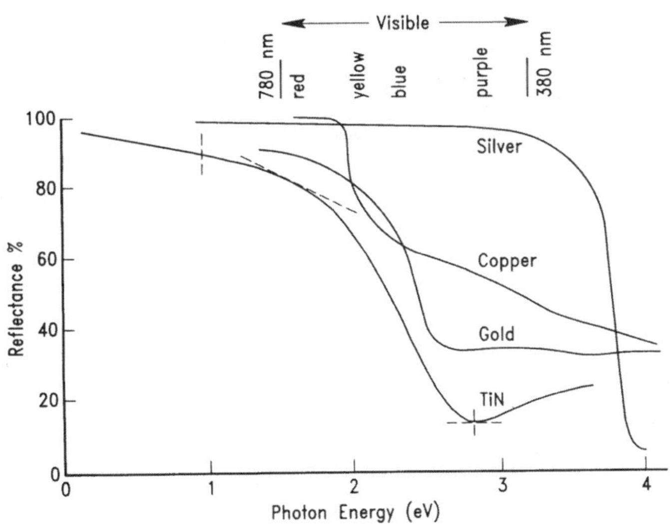

Figure 6.7   A typical reflectance curve of TiN compared with the noble metals Au, Ag, and Cu.

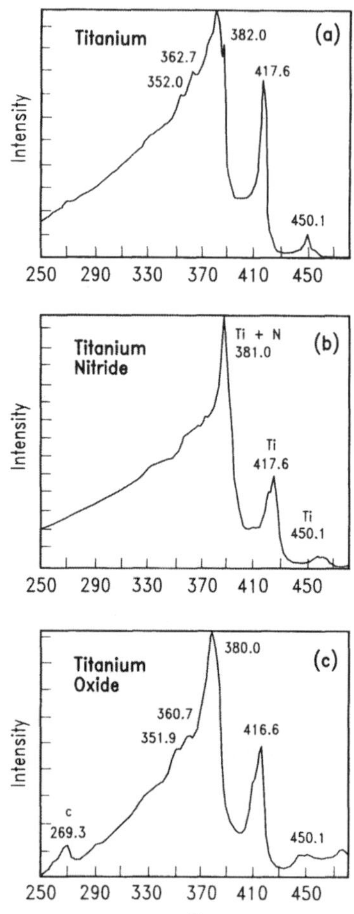

Figure 6.8   AES nondifferentiated spectra characteristic of (*a*) Ti, (*b*) TiN, and (*c*) TiO$_2$.

measured using sheet resistance, RBS, or cross-sectional SEM and TEM techniques, as described earlier in this chapter.

Microstructure   TiN$_x$ can exist over a range of composition where *x* varies from 0.6 to 1.2. Although the lattice parameter can be a good indication of stoichiometry, since it increases with nitrogen concentration, it is also influenced by defects present in the film, such as interstitials and vacancies and their annihilation during subsequent heat treatment. Characteristic AES spectra from Ti and TiN are shown in Figure 6.8. The Auger line shapes for TiN observed at 380, 417, and 480 eV are different from those observed for Ti. Determination of the Ti to N ratio is difficult in TiN, due to the overlap of the Ti and N Auger electron transitions at 380 eV. However, Auger electron characteristic peak shapes can be effectively used to discern different compounds of Ti from TiN. Sheet resistance is lowest for stoichiometric films, but a direct correlation of sheet resistance with stoichiometry is difficult because it is affected by the defects and thin-film microstructure.

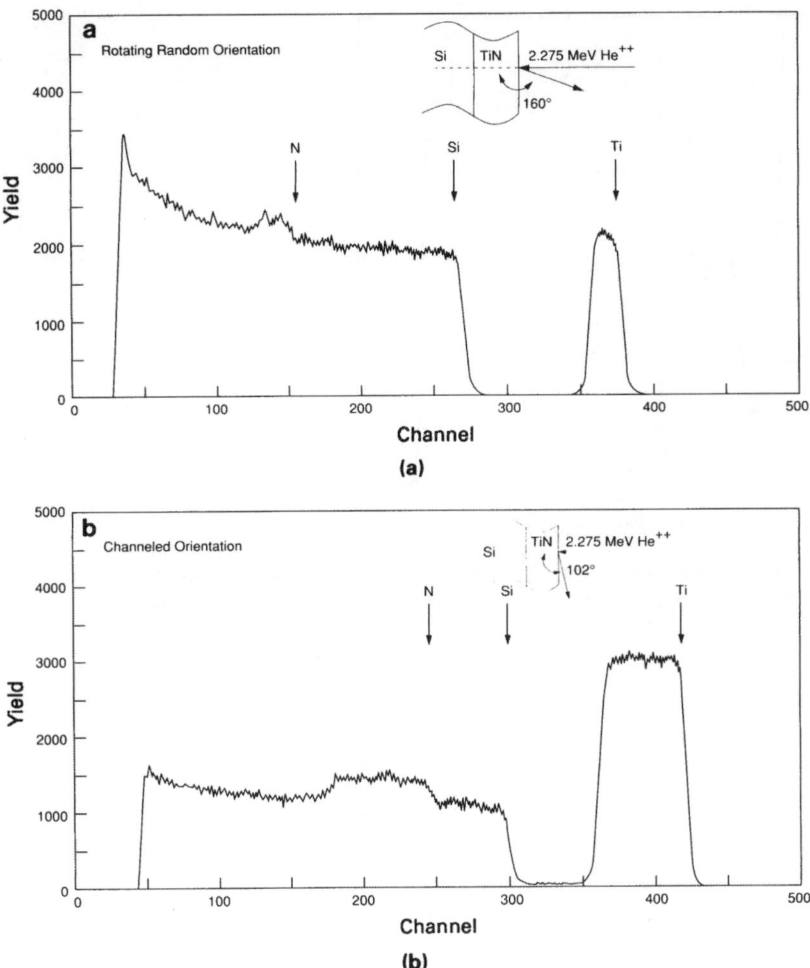

Figure 6.9 RBS spectra from 1200-Å-thick reactively sputtered TiN film on Si acquired using (a) rotating random orientation/normal detector angle and (b) substrate channeled orientation/glancing detector angle.

The most commonly used technique for TiN stoichiometry measurement is RBS. An RBS spectrum of TiN on Si, acquired with the substrate in a random orientation, is shown in Figure 6.9*a*. The signal from N is superimposed on Si due to its low atomic number compared to the substrate silicon. The concentration of Ti and N is obtained by integrating the areas under the respective peaks. The accuracy in determining the N concentration is, however, very poor due to the large background from silicon and the low RBS yield for N. The accuracy of N detection can be improved both by reducing the Si background and by using a channeling orientation (i.e., He$^+$ beam aligned with a crystallographic direction of the single-crystal Si substrate). An RBS spectrum of TiN is shown in Figure 6.9*a* acquired

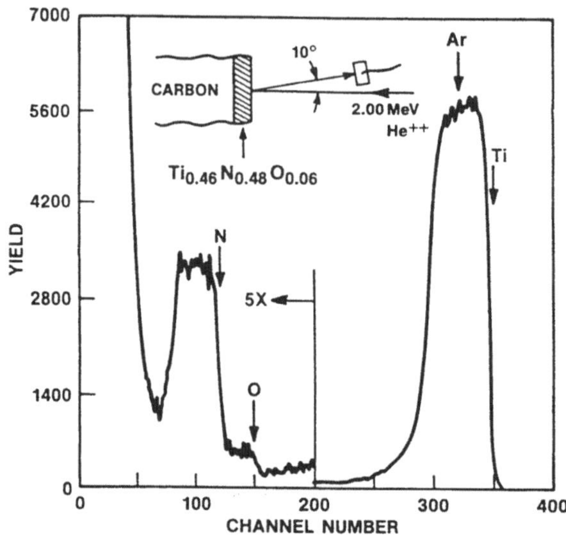

Figure 6.10 RBS spectra from a 1800-Å-thick compound target sputtered TiN film on amorphous carbon, illustrating the very low background at the N peak position.

with the substrate in a random orientation. It can be compared to the same spectrum taken in a channeled orientation using glancing angle detection, as shown in Figure 6.9*b*. The N signal in Figure 6.9*b* is better defined due to the reduced background from the substrate, and N and Ti peaks show increased energy width due to the increase in path length of the He$^+$ beam in the sample at grazing detection angle. The stoichiometry of the film is arrived at by integrating the areas under the Ti and N peaks and taking the ratio after normalizing by the scattering cross sections for the respective elements. The total error in determining the stoichiometry this way is about 15%. The stoichiometry of TiN film on Si can be measured with this accuracy up to a thickness of about 2000 Å. The RBS stoichiometry of TiN can be improved further by depositing it on a vitreous polished carbon substrate, whose atomic number is less than nitrogen. An RBS spectrum of a 1800-Å-thick TiN film on amorphous carbon acquired at a backscattering angle[27] of 170° is shown in Figure 6.10. The area under the N peak can be measured accurately because there is now no interference from the background since the signal from the carbon substrate lies to the left of the N. The total error in determining the TiN stoichiometry in this case is about 5%, but remember that it is on a model substrate (carbon) rather than on Si. Thickness measurements of the film are arrived at by taking the ratio of the areal density (the integrated N area in the RBS) to an assumed bulk density. Because of the porosity present in this film, as observed from microstructure studies, the assumed bulk density is inappropriate, and therefore the physical thickness of the film can be underestimated by as much as 50%. Hence, thickness calibration of the RBS spectrum has to be carried out using analytical

                                                      BARRIER FILMS   Chapter 6

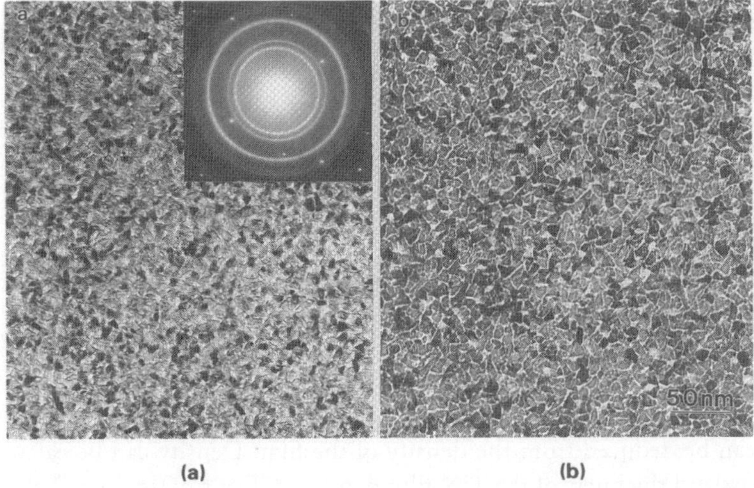

Figure 6.11 Planar TEM bright field images of TiN films processed at 4-kW power in 50%N$_2$ and 50%Ar ambient and imaged at different focus conditions in the TEM: (*a*) exact focus condition and (*b*) imaged at under-focus condition, where grain boundary regions appear as white areas. The inset in (*a*) is a selected area electron diffraction pattern from TiN.

techniques such as TEM or high-resolution SEM, which measure a physical thickness rather than the *mass* thickness.

The microstructure of reactively sputtered TiN films consists of very fine grains. The average grain size varies from 100 to 250 Å, depending upon the processing conditions. Microstructure examination of TiN requires imaging techniques with high resolution such as TEM and immersion lens SEM. A TEM plan-view bright field (BF) image obtained at exact focus conditions of a reactively sputtered 460-Å-thick TiN film processed at 4 kW in a 50% N$_2$/50% Ar atmosphere is shown in Figure 6.11*a*. The microstructure appears homogeneous, and phase analysis using electron diffraction (inset in Figure 6.11*a*) shows that the film is polycrystalline and it is cubic TiN. The grains present in the microstructure can be observed either by imaging the film using a portion of the (111) diffraction ring or imaging slightly away from the exact focus condition. An underfocused BF image of the TiN film exhibits well-defined grains, highlighted open grain boundary areas, and a network connecting all the grain boundary areas, as observed in Figure 6.11*b*. The width of the grain boundary areas, that is, the porosity, depends upon the process conditions. An XTEM image of the TiN film of Figure 6.11 exhibits very fine grains with columnar morphology, as shown in Figure 6.12. The presence of pores or voids observed between the columnar grains indicate loosely packed grains and are in agreement with the bright grain boundary areas observed in the TEM plan-view images. Similar results have been reported on the microstructure of TiN in the literature.[28] The average grain size of this film is about 110 Å. The extent of porosity

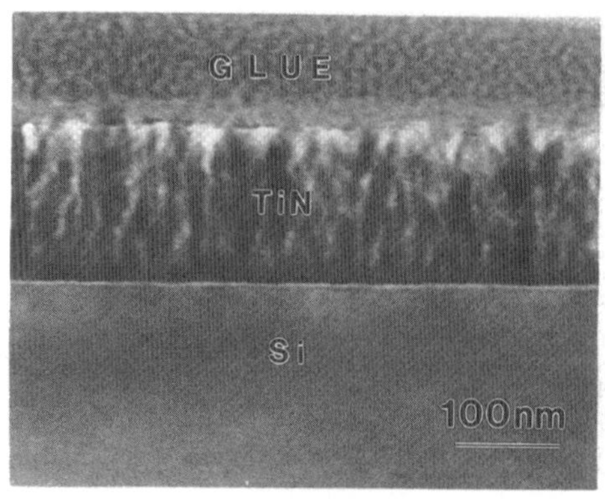

Figure 6.12 **XTEM image of TiN on Si, as for Figure 6.11, showing large regions of grain boundaries.**

in the TiN film can be deduced from the density of the film. Density can be calculated from the mass and thickness of the TiN film deposited. Density (in g/cm³) also can be calculated from the areal density of Ti or N atoms obtained from the RBS studies, from accurate determination of the physical thickness of the film, and by using the relation

$$\frac{\text{(volume density)(molecular weight)}}{\text{Avogadro's number, } 6.02 \times 10^{23})(\text{no. of atoms per molecule)}} \tag{6.2}$$

where the volume density in atoms/cubic centimeters equals the areal density in atoms/square centimeters/thickness in centimeters, molecular weight is in units of grams/mole, and Avogradro's number is in units of molecules/mole.

The density of the film shown in Figure 6.11 is about 2.42 g/cm³. This is only 42% of the bulk density, 5.22 g/cm³, of TiN, indicating a highly porous film. This calculation assumes that the film is stoichiometric. The porous film can absorb large amounts of oxygen; for example, about 16 at. % was determined to be present from RBS of the film shown in Figure 6.11. The microstructure of the TiN film can be modified by changing the processing conditions: by applying a negative bias to the substrate or by inert sputtering of a compound target. A plan-view underfocused TEM image of a reactively sputtered TiN film processed at 8 kW in a nitrogen atmosphere is shown in Figure 6.13. The extent of the grain boundary region is much smaller in this film than in the film shown in Figure 6.11. An XTEM image of this film shows that the grains are columnar and densely packed and that the grain boundaries are sharp. The density of the film calculated using RBS areal density values and the XTEM thickness is 4.88 g/cm³, which is about 87% of the bulk TiN value. The oxygen content in this film was below the RBS detection limits, which is about 5 at. %. Other common features observed in TiN films are cracks in the microstructure due to higher stress in the film when it is deposited under bias.[29]

TiN films deposited on silicon substrates are textured along either the <100> or <111> direction, depending upon processing conditions. Texture in the film is

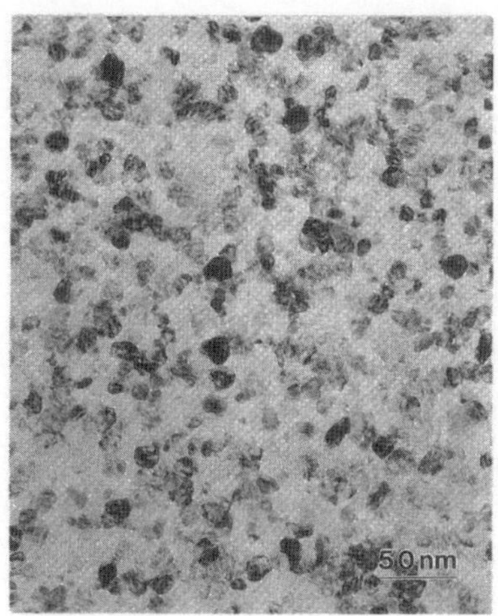

Figure 6.13   Planar TEM image of a TiN film processed at 8-kW power in a 100% N$_2$ atmosphere and imaged at underfocus conditions in the TEM.

more conducive to diffusion, due to the alignment of grains along one direction. A randomly oriented film can be produced by increasing the target power or by applying a bias to the substrate. Texture in the film is measured by using XRD. It is desirable to have a TiN film with low stress, because high film stress causes particles during processing. The stress in a reactively sputtered film is about 1×10^9 dynes/cm^2.

Barrier properties   Unlike W–Ti films, TiN films exhibit poor adhesion with Si and oxide substrates and very poor step coverage. In addition, they have high contact resistance with substrate Si. Hence, a thin layer of Ti is interposed between Si and TiN to reduce contact resistance and to improve adhesion. The effective thickness of TiN at contact bottoms can be improved either by depositing a thicker TiN film or by applying a negative bias to the substrate during sputtering, which improves the TiN step coverage. The fine columnar morphology of TiN and the difference in the mobility of sputtered species at the contacts compared to the flat surface results in weak spots at the contact corners. An XTEM of a Ti/TiN/Al–Cu/TiN(ARC) metal structure on a reflowed contact with substrate Si exhibiting breaks in the TiN film microstructure at contact corners is shown in Figure 6.14. These weak spots in the TiN potentially act as nucleation sites for interdiffusion and barrier breakdown.

Two similar devices with Ti/TiN/Al–Cu/TiN metal structures, where Al–Cu was sequentially (no air break between depositions) and nonsequentially (air break between depositions) deposited on TiN, were found to have quite different electrical properties following identical heat treatments. In the former case, device failure indicated TiN barrier breakdown, whereas normal device characteristics in the latter indicated a stable TiN barrier. Device failure is found to be caused by the formation of Al spikes in the substrate Si at the contact, resulting in junction leakage, as shown in Figure 6.15a. The Al spikes are from the faceting of etch pits along {111} planes, due to diffusion of Si into Al–Cu. An absence of any interdiffusion

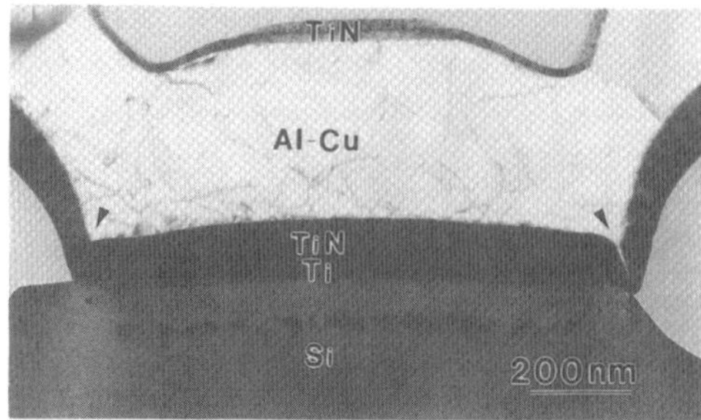

Figure 6.14 XTEM image of a reflowed contact with Ti/TiN/Al–Cu/TiN(ARC) metallization, exhibiting discontinuities in TiN step coverage at contact corners, as indicated by arrows.

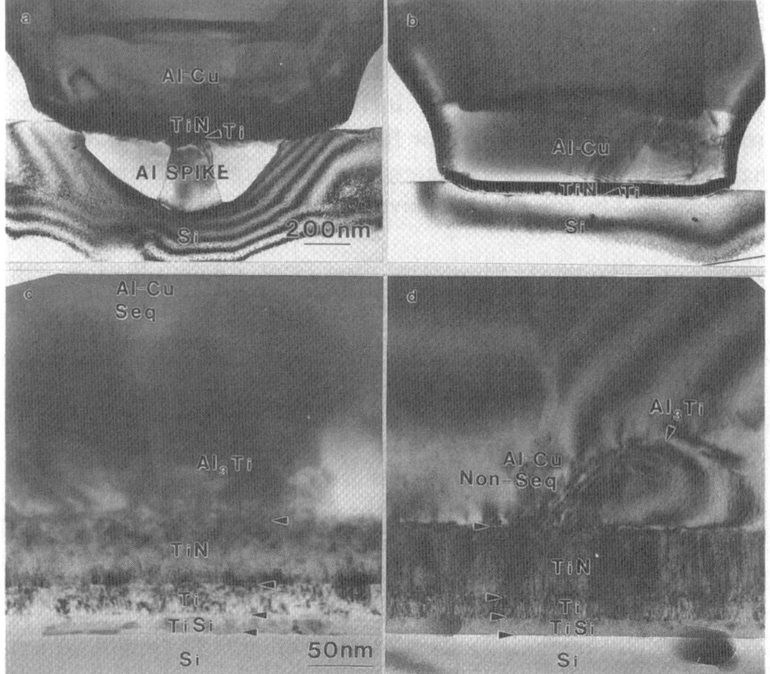

Figure 6.15 XTEM images through Ti/TiN/Al–0.5% Cu/TiN(ARC) metal structures annealed at 450 °C for 30 min: in contacts, images (*a*) and (*b*), and on blanket wafers, images (*c*) and (*d*). Al–0.5% Cu was sequentially deposited on TiN in images (*a*) and (*c*), and TiN was exposed to air prior to Al–Cu deposition in images (*b*) and (*d*). (*a*) Al spike in the substrate due to barrier breakdown and corresponding device failure. (*b*) Normal device due to robust barrier. (*c*) Intermixing between TiN and Al–0.5% Cu. (*d*) Stable TiN with sharp interface at TiN/Al–0.5% Cu. Arrows in (*c*) and (*d*) indicate interfaces corresponding to different thin film layers. The dark features at the bottom of image (*d*) are from contamination spots formed during compositional analysis using EDS in the TEM.

                                        BARRIER FILMS   Chapter 6

across TiN and an absence of reaction between TiN and Al–Cu, as shown in Figure 6.15b, produces normal device performance. Although the cause of electrical failure can be deduced from these images, a detailed study of barrier stability mechanisms is limited by the small areas of contact with Si available in a patterned wafer. This limitation can be circumvented by processing a similar metal structure on a blanket wafer and heat-treating it as if for a device wafer. The advantages of this approach include

- availability of larger areas for analysis—RBS requires large areas for analysis

- simplicity of XTEM sample preparation from a blanket Si wafer, compared to that of a device wafer

- flexibility to vary film thickness and process conditions

- simplicity in processing, since it does not require lithography for patterning.

The disadvantage is that the thickness and composition variations that result from topographies such as contacts are not duplicated on a blanket wafer. However, the barrier stability information obtained using blanket wafer study can be effectively used to develop a process with a stable barrier.

A metal structure corresponding to the contact metallization of Figure 6.15a but deposited on blanket Si and heat-treated at 450°C for 60 min is shown in Figure 6.15c. The normal columnar morphology observed in a reactively sputtered TiN film is absent and is transformed to very fine grains when the TiN and Al–Cu are sequentially deposited. Further, they have reacted to form Al_3Ti (which is confirmed by the detection of Ti within the Al–Cu layer using EDS analysis at a 1000-Å spot size on the XTEM cross-section sample). A similar XTEM image of blanket Si corresponding to the contact metallization of Figure 6.15b is shown in Figure 6.15d. The stable TiN columnar morphology and sharp TiN/Al–Cu interface observed in this figure indicate a stable TiN barrier, although some localized Al_3Ti intermetallic compound formation is observed, as indicated by the arrow in Figure 6.15d. However, these localized reactions were not commonly observed in actual device cross sections (Figure 6.15b). These comparisons demonstrate that different reactivities between the interfaces occur with and without air exposure between the depositions and that these lead to the formation of different products and microstructures. These things, in turn, lead to barrier breakdown. The large grain boundary areas in the TiN and in the Al_3Ti at the TiN/Al–Cu interface provide a pathway for Al diffusion into the Si substrate, which is followed by the diffusion of Si into the Al–Cu, resulting in barrier breakdown.

SIMS depth profiling, which is sensitive to elements down to trace levels, can also be used to monitor these diffusion effects and how they change as a function of process changes such as introducing air breaks. In general, SIMS depth profiles provide only relative concentrations, since ionization yields in the SIMS process depend on both the ion-beam source used and the material matrix being analyzed. Another problem can be the "ion-induced topography" caused by nonuniform sputtering, which can greatly degrade depth resolution at an interface and give a false impression of diffusion mixing. Hence, for the type of studies discussed here it is always preferable to make comparisons to spectra from control samples (e.g., without heat treatment, as compared to the annealed samples) to determine the extent of any changes due to reaction.

SIMS depth profiles from the above set of samples are shown in Figure 6.16. The profiles in Figures 6.16c and d were a result of using Cs+ as the primary ion source to obtain a high yield for secondary oxygen ions. Those in Figures 6.16a and b were a result of using the O^{2+} primary ion source to obtain a high yield for the secondary metallic ions. By using both sources, one can track the oxygen content with maximum sensitivity and the metallic intermixing at the TiN–Al interface at high sensitivity. A significantly higher oxygen signal is observed in the sample when the Al–Cu metal is deposited on the air-exposed barrier, compared to samples in which the metal is sequentially deposited (Figure 6.16c compared to d). These results indicate that the higher oxygen present in the TiN exposed prior to Al–Cu deposition plays a significant role in the barrier stability. Although TiN barrier stability has been attributed to the presence of oxygen in the film, there are two different mechanisms proposed for barrier stability. In one case, oxygen is believed to be adsorbed at TiN grain boundaries which react with Al, forming Al_2O_3 at the grain boundary. Any further inward diffusion of Al through the grain boundary is prohibited due to the high activation energy necessary for diffusion through Al_2O_3.[30] An alternative proposed mechanism is that the oxygen present in the TiN reacts with Al, forming a thin and continuous layer of Al_2O_3 at the TiN/Al interface, which prevents inward diffusion of Al through the barrier. There have been studies reporting the presence of a thin layer of Al_2O_3 at the TiN/Al interface supporting the latter mechanism.[5]

The oxygen uptake by TiN decreases with an increase in the density of the TiN film, owing to the decrease in the amount of grain boundary regions. For example, the amount of oxygen is about 16% in a TiN film that is only 42% dense, whereas no oxygen was detected using RBS in a TiN film which is 85% dense. The higher amount of oxygen in a porous TiN film is essential to obtaining a robust barrier properties in spite of some localized intermetallic compound formation (Figure 6.14b). The localized reaction takes place due to the presence of compositional inhomogeneities, for example, a Ti-rich spot. These spots act as nucleation spots for barrier breakdown when there is insufficient oxygen present in TiN to stabilize the barrier, for example, in dense barrier films. Further, the out-diffusion of N from N-rich films into Al during the postmetallization treatments could affect the reliability of the metal during its lifetime. This can be studied by SIMS depth profiling samples with and without postmetallization treatments. An additional heat treatment of TiN at 400°C prior to metal deposition has been found to cure the weak spots and stabilize the barrier.[31] The choice of TiN film processing conditions,

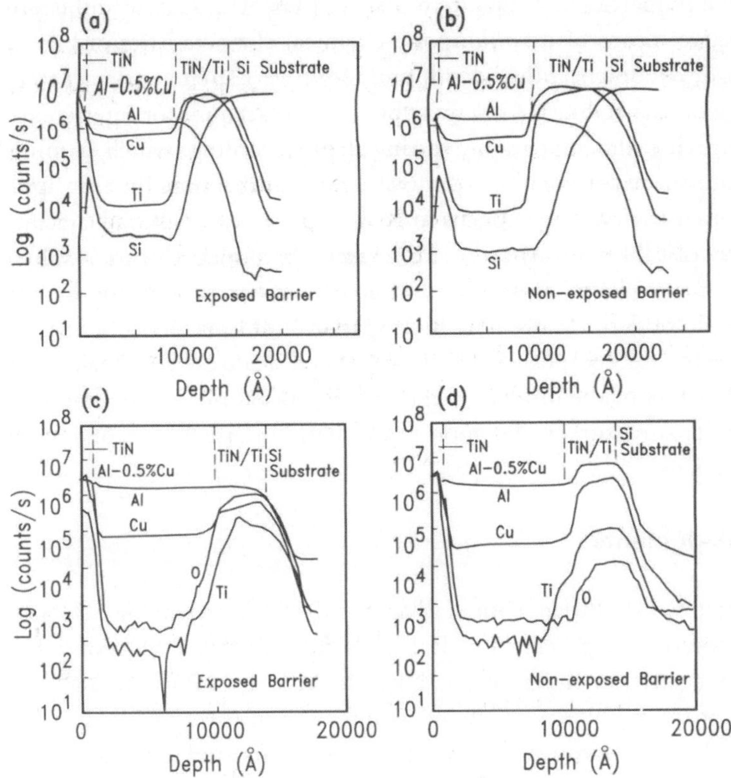

Figure 6.16   SIMS depth profiles through Al–0.5% Cu/TiN/Ti/Si samples annealed at 450°C for 30 min using an oxygen source: (*a*) TiN barrier exposed to atmosphere prior to Al–0.5% Cu deposition and (*b*) Al–0.5% Cu sequentially deposited on TiN. Similar depth profiles obtained using a cesium primary source are shown (*c*) for exposed barrier and (*d*) for unexposed barrier. Note that the oxygen concentration is higher in the exposed barrier film than in the unexposed film.

thickness, and barrier treatment has to be made by taking into account the throughput, time, and postmetallization heat-treatment temperatures.

6.6 Summary

It has been established that barrier films must fulfill stringent and conflicting requirements for integration into a metallization with robust properties. For example, a barrier film has to remain chemically inert with the interconnect Al metal and, at the same time, form a low-resistance contact with the substrate, which requires chemical interaction. The role of analytical techniques in characterizing the film microstructure, developing a robust barrier film, and monitoring the film during processing has been clearly established by the case histories discussed here. Depth profiling and imaging, combined with localized compositional analysis, is essential for characterizing the properties and stability of barrier films AES surface and depth profiling is an effective means of examining the chemical characteristics of barrier films during barrier development. SIMS depth profiling is very useful to detect trace elements to examine barrier stability. Care must be taken to account for interference from ion induced mixing and topography during depth profiling, which is minimized by carrying out the experiment on a control sample at the same time. In spite of its only moderate sensitivity, RBS happens to be one of the best tools to characterize the stoichiometry of TiN films. Appropriate corrections to mass thickness values obtained from RBS have to be made to account for the porosity observed in barrier films and to give a physical thickness. The high spatial resolution of the TEM for imaging is also available for compositional and structural analysis. This makes it a valuable tool for the study of the properties and stability of barrier films because of their fine-grained microstructure and complexities of the intermetallic compound formed during processing.

Acknowledgments

The author would like to acknowledge the staff at the Materials Technology Department of Intel Corporation, especially Dr. P. Davies for valuable discussions; Dr. B. Tracy, Mr. M. Frost, Mr. L. Kulig, Ms. P. Nielson, Ms. H. Humiston, and Ms. C. Matos for some of the analysis support and results used in this chapter; and Dr. R. McDonald and Dr. J. Mardinly for valuable comments and critical review. Acknowledgments are also due to Dr. S. Bauman at Charles Evans and Associates and Dr. J. Fair at Varian Associates for very useful discussions and critical review.

References

1   C. R. M. Grovenor. *Microelectronic Materials*. Adam Hilgar, Bristol, 1989, p. 270.

2   R. S. Nowicki, J. M. Harris, M.-A. Nicolet, and I. V. Mitchell. *Thin Solid Films*. **53**, 195, 1978.

3   R. S. Nowicki and I. Wang. *Thin Solid Films*. **15**, 235, 1978.

4   R. S. Nowicki and M.-A. Nicolet. *Thin Solid Films*. **96**, 317, 1982.

5   J. M. E. Harper, S. E. Hornstrom, O. Thomas, and A. Charai. *J. Vac. Sci. Technol.* **A7**(3), 875, 1989.

6   M.-A. Nicolet. *Thin Solid Films*. **52**, 415, 1978.

7   R. S. Nowicki and B. Schiefelbein. "Tungsten and Other Refractory Metals for VLSI Applications." *Proceedings*. Mat. Res. Soc, 1986, p. 341.

8   M. Wittner. *J Vac. Sci. Technol.* **A3** (4), 1787, 1985.

9   A. J. Aronson. *Microelectronic Manufacturing and Testing*. 1988, p. 506.

10   M. Wittner. *J. App. Phys.* **53** (2), 1007, 1982.

11   A. Sherman and I Raaijmakers. *Eleventh International Conference on CVD 1990.* (K. E. Spear and G. W. Cullen, Eds.) The Electrochemical Society, Pennington, NJ, 1990, p. 374.

12   I. Raaijmakers, R. N. Vrtis, J. Yang, S. Ramaswami, A. Lagendijk, D. A. Roberts, and E. K. Broadbent. *Proceedings.* Mat. Res. Soc. Symp., Vol. 260, 1992, p. 99.

13   G. J. P. Krooshof, F. H. P. M. Habraken, and W. F. van der Weg. *J. Appl. phys.* **63** (10), 5104, 1988.

14   S. Gupta, J.-S. Song, and V. Ramachandran. *Semiconductor International.* Oct. 1989, p. 80.

15   *CRC Handbook of Chemistry and Physics.* 65th ed, CRC Press, Boca Raton, FL, 1984–85, p. B-154.

16   JCPDS X-Ray Diffraction File Card No. 5–582.

17   R. W. Bower. *Appl. Phys. Lett.* **23** (2), 99–101, 1973.

18   M. Eizenberg, K. N. Tu, C. J. Palmstrom, and J. W. Mayer. *Appl. Phys. Lett.* **45** (9), 905–907, 1984.

19   B. W. Shen, J. M. Anthony, P.-H. Chang, J. Keeman, R. Matyi, and H. L. Tsai. *Mat. Res. Soc. Symp Proc.* **54**, 103–108, 1986.

20   I. Krafcsik, J. Gyulai, C. J. Palmstrom, and J. W. Mayer. *Appl. Phys. Lett.* **43** (111), 1015–1017, 1983.

21   A. G. Dirks, R. A. M. Wolters, and A. E. M. De Veirman. *Mat. Res. Soc. Symp. Proc.* **260**, 787–793, 1992.

22   H. E. Bishop. In *Methods of Surface Analysis Techniques and Applications.* (J. M. Wells, Ed.) Cambridge University Press, 1989, p. 87.

23   J. O. Olowolafe, C. J. Palmstrom, E. G. Colgan and J. W. Mayer. *J. Appl. Phys.* **58** (9), 3440–3448, 1985.

24   A. S. Bhansali, I. J. M. M. Raaijmakers, R. Sinclair, A. E. Morgan, B. J. Burrow, and M. Arst. *Mat. Res. Soc. Symp. Proc.*

25   JCPDS X-Ray Diffraction Card File No. 6–0642.

26   A. J. Perry. *J Vac. Sci. Tech.* **A6** (3), 2140–2214, 1988.

27   C.-S. Wei, M. L. A. Dass, T. Brat, and D. B. Fraser. "Tungsten and Other Refractory Metals for VLSI Applications IV" *Proceedings.* Mat. Res. Soc, 1989, pp. 283–288.

28   I. Petrov, L. Hultman, J. E. Sundgren, and J. E. Greene. *J. Vac. Sci. Technol. A.* **10** (2), 265–272, 1992.

29   N. Kumar, J. T. McGinn, K. Pourrezaei, B. Lee, and E. C. Douglas. *J Vac. Sci. Tech.* **6** (6), 1602–1608, 1988.

30 W. Sinke, G. P. A. Frijilink, and F. W. Saris. *Appl. Phys. Lett.* **47** (5), 471–473, 1985.

31 J. Hems. *Semiconductor International.* 100–102, Nov. 1990.

Appendix: Technique Summaries

The technique summaries in the following pages of this appendix include many one page summaries marked with an asterisk. These techniques are fully described in the *Encyclopedia of Materials Characterization* by C. Richard Brundle, Charles A. Evans, Jr., and Shaun Wilson. The remaining summaries are for techniques which do not appear in that volume.

Auger Electron Spectroscopy (AES)* **1**

In Auger Electron Spectroscopy (AES), a focused beam of electrons (typically a few keV to a few 10's of keV in energy) strikes the sample, causing electrons to be ejected. Some of these (the Auger electrons, named after Pierre Auger who first observed them) have kinetic energies characteristic of the atoms from which they came and so identification of the elements present in the sample can be made directly from the measurement of these energies. On a finer scale, it is sometimes also possible to determine the chemical state of the element from small shifts in the Auger energies caused by the chemical bonding of the element (chemical shift). For a solid, AES probes 2–20 atomic layers deep, depending primarily on the kinetic energy of the ejected Auger electron concerned, and somewhat on the material. With appropriate standards the relative concentration of the elements present within the probing depth can be estimated from the relative intensities of the Auger peaks in the spectrum. The great strength of AES is the high spatial resolution achievable through use of a focused beam (down to 10's of nm), and the ability to combine measurement with ion sputtering material removal, to obtain a three-dimensional elemental profile. The main use is with metals and conducting or semiconducting inorganic materials, since beam damage and charging can be issues with non-conducting and organic materials. Many Auger spectrometers are designed to maximize sensitivity (typically down to 100 ppm of element concentration) and speed of analysis, and do not have the spectral resolution to use chemical shifts for chemical state identification. Auger is capable of identifying all elements, except hydrogen and helium.

Ballistic Electron Emission Microscopy (BEEM)   2

PHILLIP NIEDERMANN

Ballistic electron emission microscopy (BEEM) is a technique used to investigate the electronic structure of buried interfaces with high spatial resolution. Primarily, BEEM can be used to locally measure and produce images of the height of subsurface electronic barriers, such as Schottky barriers on n- or p-doped semiconductors. In addition to that, interface electronic state density, hot carrier transport, and doping profiles can be investigated by BEEM. The basic measuring instrument is a scanning tunneling microscope (STM) used in a three-terminal configuration with the tip as electron emitter, a base through which electrons are transmitted without scattering (hence the term ballistic), and a collector separated from the base by an electronic barrier (the specimen), such as a Schottky barrier. The spatial resolution depends on the carrier scattering properties of the base; it can be as high as ~1 nm. Since the BEEM image of an interface is taken while the STM tip is scanning the surface, an atomic-resolution topographic image of the surface can be obtained simultaneously with the interface image.

Information	Spatially resolved barrier height for Schottky barriers, heterojunctions, p–n junctions, etc. Interface electronic state density. Hot carrier scattering. Recombination/generation currents
Destructive	No
Environment	Vacuum, inert atmosphere or air
Sample requirements	Schottky contact or other electronic barrier under a conducting layer thinner than a few tens of nm. Zero-voltage impedance across the barrier >100 kΩ (for low barriers, small device area and/or low temperature required)
Lateral resolution	Down to 1 nm
Imaging	Inherent in technique
Measuring equipment	STM
Cost	As for an STM, $50,000–$50,000

Introduction

For surface analysis, a broad range of well-established techniques exist for the study of electronic and structural properties of surfaces and thin overlayers. Interfaces between bulk materials, though important for electronic and other material properties, are much more difficult to access with microscopic or spectroscopic methods. The measurement of electronic barrier heights with submicrometer resolution was not possible before BEEM was invented.

As with STM, BEEM[1] can be performed in air, ultra-high vacuum, or liquids at temperatures ranging from cryogenic to above room temperature. In general terms, BEEM is a method to inject hot carriers with controllable energy into a solid and to measure their collection efficiency directly or via a secondary process. There is an entire class of BEEM-type experiments, including ballistic hole emission microscopy, interface band structure spectroscopy, and electron-electron scattering spectroscopy. Since it is a very recent development, with much exploratory research currently underway, BEEM will be further expanded to include even more variations and uses in the future.

Experimental Setup

The experimental setup of BEEM is shown in Figure 1 for a Schottky barrier on an *n*-doped semiconductor. A constant current passes, via tunneling, from the tip of an STM into the sample, regulated by the feedback circuitry as in a standard STM imaging mode. The sample is a layered system having at least two layers separated by an electronic barrier. Until now, most measurements have been done on *n*-type or *p*-type Schottky barriers, with the metal layer ("base") grounded and the semiconductor ("collector") held at virtual ground and connected to an electrometer amplifier. Electrons that tunnel into the base may travel without energy loss to the metal–semiconductor interface. If their energy is high enough, they may enter the semiconductor and drift away from the interface under the influence of the electric field in the depletion zone of the semiconductor. A current-voltage converter detects the collector current, which is typically on the order of a few picoamperes and can be measured either as a function of tunnel voltage so as to

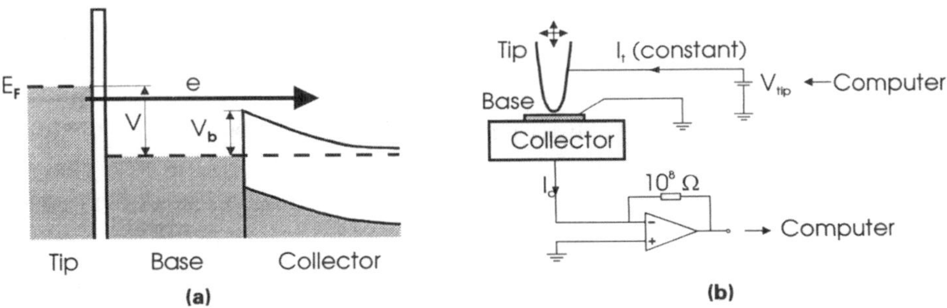

Figure 1   (*a*) **Energy level diagram for BEEM process,** (*b*) **Schematic of experimental setup.**

obtain BEEM spectra or as a function of position at a constant tunnel voltage so as to obtain interface images.

Physical Principle

The transmission of an electron can be described in three steps:

1 Tunneling from tip to base. This produces a density of hot electrons in the base that decreases exponentially below the Fermi energy of the tip and away from the surface normal.

2 Transport through the base. If the base is thin, as compared to the elastic mean free path, a large fraction of the electrons can reach the interface without scattering. In gold, for example, the mean free path is about 130 Å,[2] and a base layer of 100 Å is adequate for a typical BEEM measurement with a tunnel current on the order of 1 nA.

3 Transmission across the interface. For an ideal interface, transmission can occur if an electronic state in the semiconductor is available with the same energy and the same **k** vector component parallel to the interface ($k_\parallel$) as the impinging electron. If the kinetic energy of the electron is just barely above the conduction band minimum of the semiconductor, $k_\parallel$ conservation dictates that only those electrons whose **k** vector lies in a narrow cone in the base can be transmitted.

A model for the collector current in BEEM has been given by Bell and Kaiser (BK model).[3] In this model, the tunneling between tip and base is described by the transmission probability $D(E_x)$ as a function of tip normal kinetic energy E_x, assuming a planar geometry.

The onset of the collector current, as predicted from this model, is quadratic, in agreement with experiment. The local barrier height can therefore be determined to a good approximation by plotting dI_c/dV versus V (see Figure 1) and performing a linear fit in the onset region.

Among the physical effects that should be taken into account in addition to those in this model are elastic scattering in the base[4] and quantum mechanical reflection at the interface.[5]

Sample Requirements

Samples that can be studied are Schottky barriers or other barriers covered by a thin metallic layer of a thickness comparable or lower than the mean free path of electrons in the eV range—typically on the order of 100 Å. If the base layer is an inert material such as gold, BEEM can be performed in air, rather than in ultra-high vacuum or inert atmosphere. The barrier impedance should be 100 kΩ or higher, requiring Schottky diodes with a barrier lower than about 0.7 eV to be cooled to a low temperature (77 K) for BEEM measurements (assuming an area of 0.1 cm^2).

Spatial Resolution

The BK model predicts a very high spatial resolution for a semiconductor with a conduction band minimum at $k_\parallel = 0$. A small critical angle above threshold implies that interface changes on a scale much smaller than the thickness of the base are measurable, putting the spatial resolution of the technique in the nanometer range. This is an ideal case; in practice, the resolution may be only in the 10 nm range for a variety of possible reasons:

1 The conduction band minimum is not at $k_\parallel = 0$, as is the case for Si(111)
2 Multiple elastic scattering occurs in the base before an electron reaches the interface; in this case, the grain structure of the base may strongly influence the resolution
3 Electrons are scattered at the interface, relaxing the conservation of $k_\parallel$. This is expected, in particular, for a nonepitaxial interface.

The attainable resolution is thus dependent on the nature and quality of the base layer and is subject to ongoing research.

In Schottky barriers, variations of barrier height at the interface are screened in the semiconductor. A small spot with a reduced barrier may be "pinched off," with a saddlepoint occurring in the potential barrier.[6] The lower the electric field near the interface, the more pronounced this effect.

Related Techniques

As already mentioned, BEEM is a general-purpose technique to inject low-energy electron or hole beams into a layered solid system; a number of BEEM-related techniques have already been demonstrated. Schottky barrier heights on p-doped semiconductors have been successfully measured[7] at 77 K using ballistic hole spectroscopy, in which hot holes are injected into the base by positively biasing the tip, in direct analogy to electron BEEM.

If the tip is biased *positively* on an n-doped Schottky diode, a collector current can again be observed that has the same sign as the BEEM current. This "reverse BEEM" current is created by the decay of hot holes in the base via collisions with electrons[8]—in analogy to an Auger process—and provides a tool for the study of carrier scattering in the metal. A similar experiment has also been demonstrated on p-doped Schottky diodes.[8]

Applications

BEEM has already been used to characterize Schottky barrier height and its spatial homogeneity in a variety of metal/n-type semiconductor systems, such as Au/Si(100), on which the initial BEEM work was done.[1]

The PtSi/n-Si(100) system[9, 10] is a sharp, easy-to-form interface that has important applications because of its high Schottky barrier of ~0.90 eV. It has been studied by BEEM for a range of silicide thicknesses.[9] The silicide layer on these diodes was found to have a relatively complex morphology, containing granular regions of local epitaxy. This granularity is reflected in BEEM images (Figure 2). For this system, the spectral behavior depended on whether samples had been exposed to air or had been produced and studied entirely in situ. In the former case, the spectra showed considerable variations in threshold and intensity, probably due

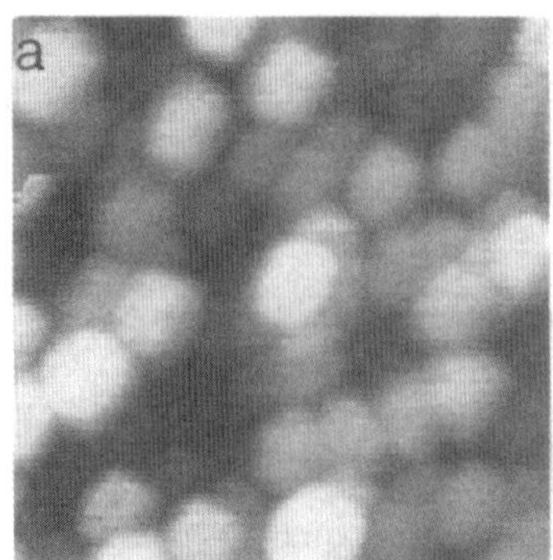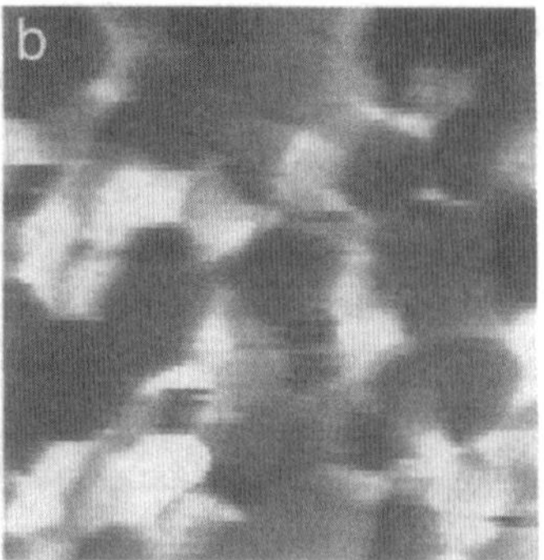

Figure 2 (*a*) STM topograph and (*b*) collector current image taken on a sample with 50 Å of PtSi on *n*-Si(100) showing strong variations of the BEEM current, which are correlated to the grain structure of the layer. Scales from black to white are (*a*) 40 Å and (*b*) 57 pA. Tunnel voltage and current were 1.5 V and 1 nA, respectively.

to surface impurities (Figure 3). On the other hand, for the in situ produced samples, the spectral onsets were constant as a function of position while the intensity still varied strongly (up to two orders of magnitude) over a scale as small as 100 Å. The variations in intensity are closely related to the granular topography and are probably due to different morphologies of the interfaces on differently oriented grains. The average collector current intensity was about an order of magnitude higher than for the ex situ samples of the same thickness, indicating a longer electron mean free path for the former.

Ion bombardment of silicon is known to introduce *n*-type defects. For a Schottky barrier, this is believed to create a high electric field near the interface, which in turn reduces the effective barrier height through the effect of the image force and electron tunneling. A BEEM study[10] on PtSi/*n*-Si(100) diodes subjected to a relatively strong rf plasma treatment showed the Schottky barrier to be lowered (Figure 3), in reasonable agreement with diode *I–V* measurements, indicating that the barrier lowering occurred uniformly. Modeling the BEEM spectra for a high-field barrier showed that a sizable fraction of the BEEM current was due to tunneling across the interfacial barrier. If the interface electric field is high, the "pinch-off" effect mentioned previously is strongly reduced, and intentionally introducing high doping should allow one to measure Schottky barrier variations on a very small scale in non-uniform systems, which are probably more common than is generally believed.

Besides Schottky barriers, other types of electronic barriers can be studied by BEEM. In particular, BEEM has been used to characterize *p–n* junctions in Si.[11] Application to thin insulating (e.g., oxide) barriers should also be possible.

Another fascinating development is the modification of interfaces using the STM itself. A study has been performed in which nanoscale features have been

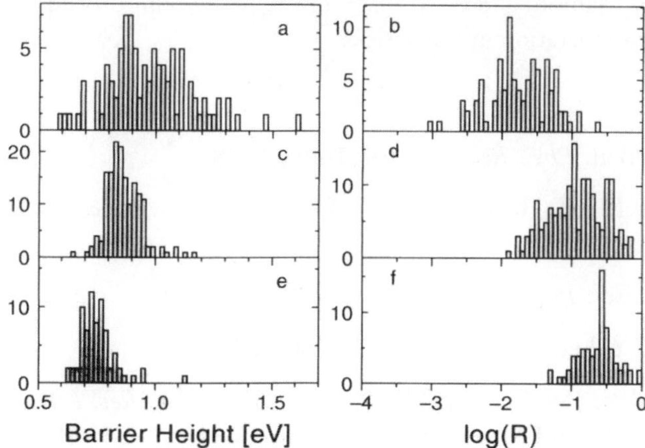

Figure 3 Histograms of barrier heights V_b and scale factors R for differently prepared PtSi/n-Si(100) Schottky diodes fabricated by Pt sputter-deposition on Si and annealing at 500 °C. In situ produced diodes (*c,d*) had a much more narrow V_b distribution and a higher R than ex situ diodes (*a,b*) (PtSi thickness 50 Å). Exposing the Si substrates to strong rf sputter discharge cleaning before producing PtSi diodes (30 Å) resulted in Schottky barrier lowering (*e,f*).

written and imaged by BEEM on an Au-Si(100) interface.[12] This opens the possibility of subsurface nanofabrication.

Conclusion

The ability of BEEM to locally measure Schottky barrier heights is now well demonstrated for a variety of materials, showing generally good agreement with other types of measurements. Beyond barrier height and interface electronic structure, BEEM can be used to investigate hot carrier transport in metal films, making this technique useful for studying film growth properties with nanometer spatial resolution.

The future will bring more uses of BEEM and related techniques as more is learned on hot carrier transport physics in metals and semiconductors, Schottky barrier local structure, and interface electronic structure. The study of fully epi-taxially grown systems will deepen our understanding of the ballistic transmission process. The realm of BEEM has already been expanding and will continue to do so, from Schottky barriers on homogeneously doped semiconductors to a variety of specially designed electron barrier systems. The combination of BEEM, TEM, surface analytical tools, and other experimental techniques will deepen our understanding of specific metal–semiconductor systems.

Acknowledgments

The author would like to acknowledge the efforts of his collaborators at the University of Geneva, Lidia Quattropani and Ø. Fischer, and to thank L. D. Bell,

M. H. Hecht, W. J. Kaiser, S. Manion, and A. Milliken of the JPL group for valuable discussions and help with the preparation of this article.

References

1   W. J. Kaiser and L. D. Bell. *Phys. Rev. Lett.* **60**, 1406, 1988.

2   L. D. Bell, W. J. Kaiser, M. H. Hecht, and L. C. Davis. *J. Vac. Sci. Technol. B.* **9**, 594, 1991.

3   L. D. Bell and W. J. Kaiser. *Phys. Rev. Lett.* **61**, 2368, 1988.

4   L. J. Schowalter and E. Y. Lee. *Phys. Rev. B.* **43**, 9308, 1991.

5   M. Prietsch and R. Ludeke. *Surf. Sci.* **251/252**, 413, 1991; *Phys. Rev. Lett.* **66**, 2511, 1991; R. Ludeke and M. Prietsch. *J. Vac. Sci. Technol. A.* **9**, 885, 1991.

6   J. L. Freeouf, T. N. Jackson, S. E. Laux, and J. M. Woodall. *J. Vac. Sci. Technol.* **21**, 570, 1982; R. T. Tung. *Appl. Phys. Lett.* **58**, 2821, 1991.

7   M. H. Hecht, L. D. Bell, W. J. Kaiser, and L. C. Davis. *Phys. Rev. B.* **42**, 7663, 1990.

8   L. D. Bell, M. H. Hecht, W. J. Kaiser, and L. C. Davis. *Phys. Rev. Lett.* **64**, 2679, 1990.

9   P. Niedermann, L. Quattropani, K. Solt, A. D. Kent, and Ø. Fischer. *J. Vac. Sci. Technol. B.* **10**, 580, 1992.

10   L. Quattropani, K. Solt, P. Niedermann, I. Maggio-Aprile, Ø. Fischer, and T. Pavelka. To be published in *Appl. Surf. Sci.*

11   S. J. Manion, L. D. Bell, W. J. Kaiser, M. H. Hecht, R. W. Fathauer, A. M. Milliken, and V. Narayanamurti. To be published.

12   H. D. Hallen, A. Fernandez, T. Huang, R. A. Buhrman, and J. Silcox. *J. Vac. Sci. Technol. B.* **9**, 585, 1991.

Capacitance–Voltage (C–V) Measurements 3

GEORGE N. MARACAS

The method of measuring free carrier (or doping) concentration profiles in semiconductors by capacitance–voltage (C–V) relies on modulating the depletion region in an MOS capacitor or field effect transistor (MOSFET), metal Schottky barrier, electrolyte–semiconductor contact, or p–n junction by an applied voltage. Superposition of a small ac voltage onto the dc bias allows the measurement of a capacitance of a thin region of doped material underneath a well-defined geometry contact (i.e., in which the area is precisely known). Scanning the reverse dc bias from zero to the breakdown voltage of the doped material provides a measure of the free carrier concentration versus depth. In simple structures such as Schottky barriers and p–n junctions the extraction of the doping profile is relatively simple. If all the dopant atoms are electrically active, then this information is obtained by analyzing the C–V characteristics, where the depletion depth is proportional to the area divided by the capacitance. The doping profile is proportional to $1/[A^2 d(1/C^2)/dV]$.

In the case of an MOS capacitor[1] the measurement is more complicated, in that special care must be taken since the device can be biased in accumulation, depletion, or inversion by different combinations of dc voltage and ac signal frequency. In the MOS capacitor case, extraction of interface state density and insulator mobile charge is also possible. The latter is achieved by controlling the mobile charge location with an applied dc bias and an elevated temperature stress cycle, which increases the mobile charge diffusion rate.

Commercial C–V measurement systems have a high degree of automation for material parameter extraction, including automatic bias scans and high temperature cycles. Contacting schemes have metal probes to contact pads, chemical contacts, and liquid metal contacts. The liquid metal contacting method is the mercury probe,[2] where liquid mercury produces a rectifying contact to the surface, obviating contact metallization patterning. Electrolytic C–V profilers[3] are available in which the rectifying barrier is produced by an electrolytic solution and the capacitance is measured at a constant dc voltage. The semiconductor is automatically and controllably etched to obtain depth information; thus, the limitation of depth from the breakdown voltage for a particular doping is relaxed.

Information	Free carrier concentration
Range of elements	Not element specific
Destructive	Yes, Hg probe no
Depth profiling	Yes
Depth resolution	Limited by Debye length

Depth limit   C–V breakdown voltage depth; electrolytic C–V no limitation

Doping range   ~10^{14}–10^{19} cm^3

Sample requirements   Depends on contacting scheme

Cost $5000–$50,000 (depends on contacting scheme and automation level)

References

1 D. K. Schroder. *Semiconductor Material and Device Characterization*. Wiley-Interscience, New York, 1990.

2 P. S. Schaffer and T. R. Lally. *Solid State Tech.* **26**, 229–233, April 1983.

3 P. Blood. *Semi. Sci. Tech.* **1**, 7–27, 1986.

Deep Level Transient Spectroscopy (DLTS)   4

N. M. JOHNSON

Deep level transient spectroscopy (DLTS) is an experimental technique for characterizing the electronic properties of deep level defects in semiconductors. It is the most sensitive spectroscopy yet devised for the detection and characterization of such defects. The measurement is performed on semiconductor devices that possess a voltage-modulable space charge layer. The most commonly used devices are Schottky-barrier diodes, p–n junction diodes, and metal–insulator–semiconductor (MIS) or metal-oxide semiconductor (MOS) capacitors. For a given deep level defect, the DLTS measurement can, in principle, yield several phenomenological parameters that characterize the defect and its interaction with free charge carriers. These parameters include activation energies for emission of charge carriers, cross sections for the capture of free carriers, and the spatial (depth) distribution of the defect, the last parameter being a property of the material rather than of the defect. In combination with perturbation techniques (e.g., uniaxial stress) or complementary measurements (e.g., electron spin resonance), DLTS can be used for the microscopic identification of electronic defects.

Main Use	Electronic characterization of deep level defects in semiconductors
Sample requirements	Electronic devices with voltage-modulable space charge layers (e.g., Schottky diode)
Quantification	Activation energies for charge emission, cross sections for carrier capture, and defect concentrations
Sensitivity	Typically 10^{-4} of the doping concentration (e.g., 10^{11} defects/cm^3 for 10^{15} dopants/cm^3)
Depth probed	Space charge layer of device, upper limit set by reverse bias breakdown (e.g., ≤ 20 μm for 10^{15} dopants/cm^3 in Si)
Instrument cost	\$50,000–\$120,000
Size	Standard laboratory bench or equivalent floor space

Recommended Reading

Johnson, N. M. "Measurement of Semiconductor-Insulator Interface States by Constant-Capacitance Deep Level Transient Spectroscopy." *J. Vac. Sci. Tech.* **21** (2), 303–314, 1982. A review of the DLTS measurement and analysis for the determination of interface state distributions in MIS capacitors.

Johnson, N. M., D. J. Bartelink, R. B. Gold, and J. F. Gibbons. "Constant-Capacitance DLTS Measurement of Defect Density Profiles in Semiconductors." *J. Appl. Phys.* **50** (7), 4828–4833, 1979. Describes the constant-capacitance mode of the DLTS measurement and its formulation for determining depth profiles of deep levels.

Lang, D. V. "Fast Capacitance Transient Apparatus: Application to ZnO and O centers in GaP *p–n* Junctions." *J. Appl. Phys.* **45** (7), 3014–3022, 1974. The first of the two original articles introducing the DLTS technique.

Lang, D. V. "Deep Level Transient Spectroscopy: A New Method to Characterize Traps in Semiconductors." *J. Appl. Phys.* **45** (7), 3023–3032, 1974. The second of the two original articles introducing the DLTS technique.

Lefevre, H., and M. Schulz. "Double Correlation Technique (DDLTS) for the Analysis of Deep Level Profiles in Semiconductors." *Appl. Phys.* **12** (1), 45–53, 1977. Introduces the double-correlation DLTS technique for determining the depth distribution of bulk deep levels in semiconductors.

Miller, G. L., D. V. Lang, and L. C. Kimerling. "Capacitance Transient Spectroscopy." *Ann. Rev. Mat. Sci.* **7**, 377–448, 1977. An early review of the DLTS technique and its application to bulk defect characterization.

Pensl, G. "Measurement Methods." In *Landolt-Börnstein: Data Tables on Impurities and Defects*. Series III, Vol. 22, Part b (O. Madelung and M. Schulz, Eds.), Springer-Verlag, Berlin/Heidelberg, 1989, Chapt. 3. A recent comprehensive compilation of electrical measurements on semiconductors, including many variations of the DLTS technique, with an extensive list of references.

Williams, R. "Determination of Deep Centers in Conducting Gallium Arsenide." *J. Appl. Phys.* **37** (9), 3411–3416, 1966. Historically interesting as the first experimental demonstration of the detection and characterization of a deep level defect with a transient capacitance measurement.

Dynamic Secondary Ion Mass Spectrometry (D-SIMS)* 5

In Secondary Ion Mass Spectrometry, SIMS, the sample is bombarded by a primary ion beam (typically in the few 100 ev to many keV energy range). Several alternative ion species are used, depending on the requirements of the analysis, which causes atoms and clusters of atoms to be sputtered from the sample surface. These atoms and clusters consist largely of neutral species, but a small fraction of both positively and negatively charged species occur (the secondary ions). These secondary ions are passed into a mass spectromter (several different types are in use), where the number of ions at each mass/charge ratio are counted. The secondary ion counts may be turned into atomic concentartions present in the sample, if (and only if), comparison to standards of the same or very similar material is available. In Dynamic SIMS, the secondary ion signal intensities are monitored as a function of sputter time (depth), allowing concentrations to be determined as a function of depth.

The strength of Dynamic SIMS is the ability to to monitor for all atoms, including hydrogen, helium, and including isotopes, over a wide dynamic range of concentrations (down to ppb to ppt levels). A lateral resolution down to below 10 nm is possible, depending on the nature of the instrument and sample (not at trace ppb levels). A depth resolution of 2–30 nm is possible, depending on material, sputter conditions, and the depth sputtered. The first few nm depth sputtered, however, is rarely quantifiable, however, owing to the initial non-equilibrium nature of the sputtering process and the lack of appropriate standards for a region which can be substantially different from the bulk material.

The major use of Dynamic SIMS is the quantitative determination of the concentration of dopants in semiconductor material, as a function of depth, over a dynamic range of 5 orders of magnitude. It is also used in a similar way for trace impurities in metals and alloys and geological materials.

Electron Beam Induced Current (EBIC) Microscopy 6

DAVID C. JOY

Charge collection scanning electron microscopy, often referred to as electron beam induced current (EBIC) microscopy, is a technique that permits the scanning electron microscope (SEM) to image individual junctions, devices, and electrically active defects in semiconductor materials. Quantitative information about parameters such as resistivity and minority carrier diffusion lengths can also be obtained.

Main Use	Determination of position and depth of junctions and electrically active defects in semiconductors
Sample requirements	All semiconductors to which ohmic contacts can be fabricated
Destructive	No, but possible damage to gate oxides from electron beam
Lateral resolution	Typically 0.5 µm—depends on beam energy
Depth probed	Up to about 10 µm in silicon at 30 keV beam energy
Depth profiling	Yes, by changing beam energy
Detection limit	Can detect doping effects at $<10^{16}/cm^3$, defects at $<10^4/cm^2$
Quantitative	Yes, can measure diffusion lengths and carrier lifetimes
Imaging	Yes, this is the standard mode of usage
Imaging/mapping	Yes
Instrument cost	See SEM
Size	See SEM

Physical Principles

When an electron beam impinges on a semiconductor, electron–hole pairs are generated, a process which requires an amount of energy e_{eh}, where typically e_{eh} is about three times the energy of the band gap—for example, e_{eh} for silicon is 3.6 eV. If we assume that all the energy deposited by the incident beam of energy E_0 is ultimately available for the generation of electron–hole pairs, then the number of carrier pairs formed, n_{eh}, will be

$$n_{eh} = E_0/e_{eh} \tag{1}$$

For example, a 10 keV beam incident on silicon could produce $10\ 000/3.6 \approx 2800$ electron–hole pairs.

In the absence of any external excitation the resistivity of a semiconductor material is high because there are few, if any, mobile charge carriers; but when a beam is turned on, each incident electron produces electron–hole pairs, and these free charge carriers produce so-called ß or beam induced conductivity in the material.[1] Because the electron and hole have opposite charges they are electrostatically attracted and will tend to drift together through the lattice, maintaining local electrical neutrality. After a short time the electron will fall back across the band gap and recombine with the hole, giving up the energy used to form the original carrier pair in the form of cathodoluminescent radiation. If, however, a potential difference is maintained across the semiconductor, then the electric field it generates will separate the electrons and holes, because the electrons will tend to move towards the positive end of the sample, while the holes will drift towards the negative end, and a current will flow which can be monitored in an external circuit (Figure 1*a*).

In practice, a more useful procedure is to generate the field internally within the semiconductor. For example, the depletion region extending a few micrometers on either side of the physical location of a *p–n* junction contains a field of several thousand volts per centimeter (Figure 1*b*). If the incident electron beam is placed on the specimen in either the *p*-type or *n*-type regions, well away from the depleted zone, then although electron–hole pairs will be generated, the material in which they are produced is electrically neutral and has no field across it, so they will

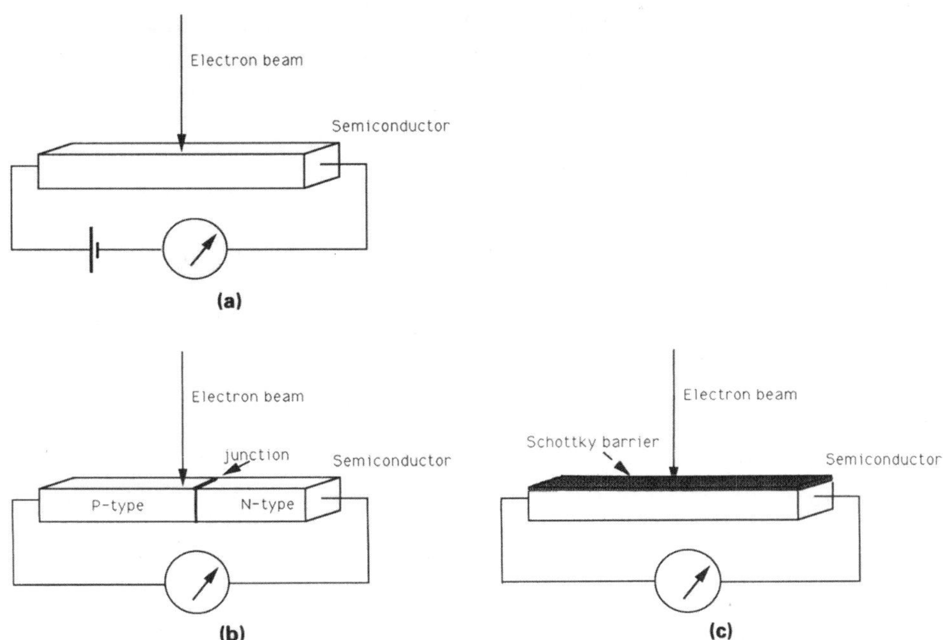

Figure 1   (*a*) ß or beam induced conductivity in a semiconductor irradiated by an electron beam. (*b*) Charge collection or electron beam induced current (EBIC) imaging with a *p–n* junction. (*c*) EBIC imaging with a Schottky barrier.

recombine and no external effect will be observed. However, if the beam is placed in the junction region, the carriers produced will see the depletion field, the carrier pairs will be separated, and the motion of these charges within the specimen will produce a flow of current I_{cc}—the charge collected or EBIC—in the external circuit given by the relation

$$I_{cc} = n_{eh} I_B = I_B E_0 / e_{eh} \tag{2}$$

since each incident electron can produce n_{eh} carrier pairs. I_{cc} is seen to be substantially greater (100× to 1000× or more) than the incident current I_B. If the beam is put close to, but not actually inside, the depleted zone, some fraction of the carriers will still diffuse back into the depletion region and be collected by the field, producing a current that will vary as $\exp(-\Delta s/\lambda)$, where Δs is the distance of the beam from the edge of the depletion zone and λ is the minority carrier diffusion length. If the incident beam of electrons is scanned, as in the SEM, then the current flowing around the loop can be amplified and displayed as a function of the beam position, producing a charge-collected, or EBIC, image. In the case of the p–n junction, the signal would show a maximum over the depletion region surrounding the junction and then fall back to zero on either side over a distance determined by the minority carrier diffusion length and the incident electron range—typically a region a few micrometers wide.

An alternative geometry (Figure 1c) is to place a Schottky barrier on the surface of a semiconductor. The depleted region extending downwards from the Schottky is again associated with a field which can separate the electron–hole pairs and given current flow in the external circuit. In this case a strong signal will be collected over the entire lateral extent of the Schottky barrier allowing, for example, for the low magnification examination of a wafer.

Practical Details

A typical imaging current, I_B, for an SEM is about 10^{-9} A at 20 keV; so from Equation 2 it can be predicted that the charge collected circuit, in silicon, would be ~0.5 μA. Measuring a current this large is certainly not a problem. However, the predicted current is actually a short-circuit current, and if the input impedance of the imaging amplifier used is R_L, then the ohmic voltage drop across the input, $I_B \cdot R_L$, is in the opposite sense to the potential at the depletion layer and, hence, will effect the signal collection efficiency. It is therefore desirable to use an amplifier with as low an input impedance (e.g., below 10 kΩ) as possible. The specimen current amplifiers used for normal SEM imaging purposes are adequate, but current-sensitive instrumentation amplifiers are superior. A signal bandwidth of 30–100 kHz is required in the amplifier for satisfactory visual rate imaging.

Electrical contacts must also be made to the material to be imaged. These contacts must be ohmic and low in resistance. This is no problem for silicon, but for some III–V and II–VI materials fabricating suitable contacts may be very difficult. As a general rule, satisfactory EBIC imaging will not be achieved unless the material

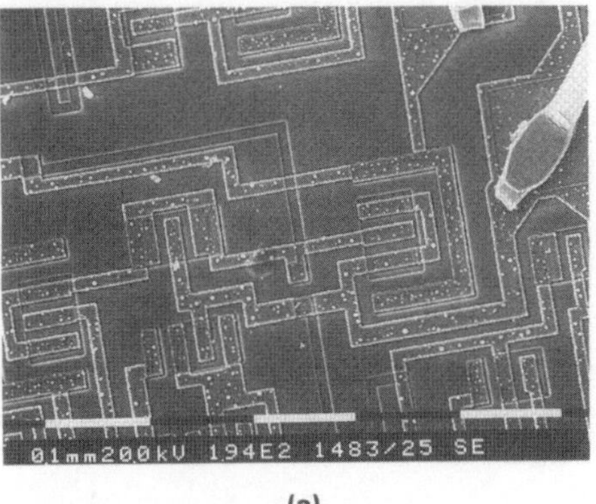
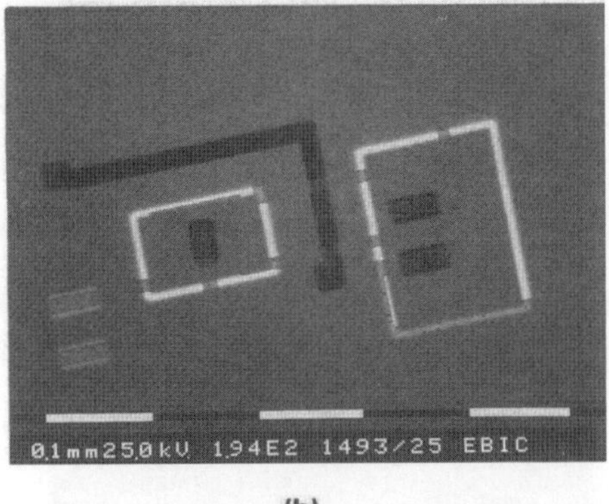

Figure 2 Secondary electron topographic image (*a*) and EBIC image (*b*) of the same area. Contrast in the EBIC image comes from current collected at junctions within the device.

or device under test shows a reverse-to-forward resistance ratio of at least ten to one when measured by a conventional volt-ohmmeter placed across the signal leads.

Applications

The original use of EBIC imaging was to observe *p–n* junctions in devices.[2] Satisfactory images can normally be obtained by collecting the signal across the $+V_e$ and $-V_e$ power rails of the device, since this will give a dc path to every junction. Compare Figure 2*a*, the normal secondary electron topographic of a device, with Figure 2*b*, the EBIC image. The junction appears as black or white, depending on whether the *p* or the *n* side of a given junction is connected to the positive-going signal lead. Junctions can be imaged at depths beneath the surface up to the maximum range of the incident electron beam—in silicon, a depth of 0.5 µm at 5 keV beam energy and 5 µm at 20 keV beam energy. If the incident beam energy is gradually increased on a given area, the energy at which a junction is first visible can be used to estimate its depth. It is not normally necessary to remove surface passivation films in order to perform EBIC imaging, but the presence of this layer will reduce the penetration of the beam into the device proper.

The Schottky barrier geometry is best suited to the study of materials in the unprocessed, or partially processed, state. In principle, the barrier regions could be as large as required to cover the area of interest, but because the capacitance represented by the Schottky diode is in parallel with the amplifier input impedance, too large a diode will give a long time-constant and require very slow scanning for a satisfactory image. It is usually better to cover the area of interest with an array of small (250–500 µm diameter) barrier dots and select the region required by moving the top contact wire to the appropriate dot. Typical contrast effects include:

1 Variations in the depletion depth due to changed resistivity (due to planned or random changes in doping) will produce corresponding changes in the collected

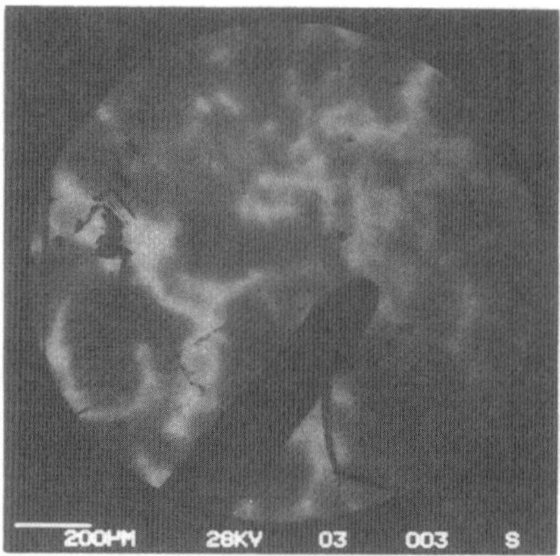

Figure 3   EBIC image of silicon wafer. The circular area shows the extent of the Schottky barrier, and the shadow is from the spring-loaded top-contact wire to the sample. The dark areas in the image correspond to regions of low depletion depth where impurities have lowered the resistivity.

signal. For example, the darker, cloud-like regions visible in Figure 3 show regions where the resistivity of a silicon wafer has been changed by impurity concentrations at the $10^{16}/cm^3$ level during crystal growth.

2 Changes in Schottky barrier height will also modulate the signal collection. For example, the A and B forms of cobalt silicide have barrier heights differing by 0.1 eV, so images of silicon formed with a mixed cobalt silicide barrier show a characteristic mottled appearance.

3 Individual electrically active defects (dislocations and stacking faults), lying within the beam range from the surface will allow the electron–hole pairs to recombine locally and produce a decrease in the collected signal.[3] Figure 4 shows such defects in a polycrystalline solar cell. The width of the defect image depends

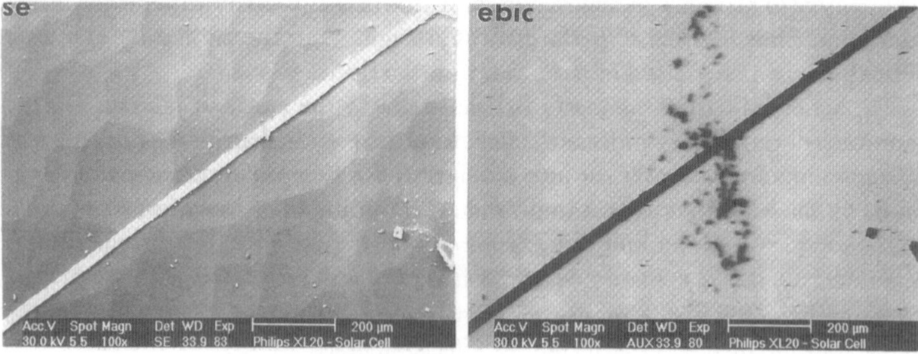

Figure 4   EBIC image of dislocations in a polycrystalline solar cell. In this case the defects are nearly end-on to the surface.

on the energy of the beam and the depth of the defect and is typically ~1–3 µm; so it is easy to see individual dislocations even in a low-magnification (100×) image. The EBIC method is therefore well-suited to the examination of materials in which the dislocation density is very low ($<10^4/cm^2$).

4 Avalanche breakdown sites at impurities may produce intense and unstable signal effects under conditions of high incident current.[4]

Although it is the imaging aspects of charge collection microscopy that have been considered here, over 90% of the published literature in this field discuss the use of EBIC to determine minority carrier diffusion lengths and lifetimes. A wide variety of techniques have been described that permit spatially resolved measurements to be performed on a 10 µm scale. For a more complete discussion, see References 4 and 5.

References

1   K. G. McKay. *Phys. Rev.* **74**, 1606–1621, 1948.

2   T. E. Everhart, O. C. Wells, and R. K. Matta. *Abs. Electrochem. Soc.* **12**, 1–5, 1963.

3   J. J. Lander, H. Schreiber, T. M. Buck, and J. R. Mathews. *Appl. Phys. Lett.* **3**, 206–208, 1963.

4   H. J. Leamy. *J. Appl. Phys.* **53**, R51–80, 1982.

5   D. B. Holt and D. C. Joy. *SEM Microcharacterization of Semiconductors.* Techniques of Physics Series, Vol. 12, Academic Press, London, 1989.

Energy-Dispersive X-Ray Spectroscopy, EDS, is a specific technique for the detection and energy distribution determination of X-Ray Fluorescence, XRF. XRF is the phenomena where X rays are emitted from a material when bombarded by high energy radiation (electrons, ions, X rays, neutrons, gamma-Rays). Some of the X-ray energies emitted are characteristic of the atoms present, allowing atomic identification in the material of interest.

In EDS a solid state X-ray detector, usually lithium drifted Si, and pulse counting electronics are used. The detector converts an incoming X-ray photon into an electronic pulse of amplitude proportional to the energy of the X ray. The signal processing electronics counts the number of pulses of each different amplitude, resulting in a histogram of X-ray energy versus intensity. The X-ray energies allow atom identification, and the relative peak intensities can be related to relative atomic concentrations by comparision to standards, or by theoretical calculations.

All elements with an atomic number higher than Li are, in principle detectable, though effectively dealing with low Z elements requires some care and correctly set up instrumentation. In practise, EDS is primarily used in conjunction with e-beam columns (SEMs, TEMs, STEMs, Auger instruments) as the excitation source. The depth probed is dependant primarily on the energy of the electron beam, and on the material being probed. It can vary from as little as 20 nm (high Z material, low beam energy) to as much as 5000 nm (low Z material, high beam energy). The lateral resolution is similar to the depth probed but is not determined by the primary beam diameter. It is determined by the beam energy and the sample material, because the beam spreads by scattering as it penetrates into the material, creating X rays in this enlarged volume.

The particular strength of EDS is the simultaneous (or parallel) detection of elements rapidly and cheaply, by placing the physically small detector inside the SEM, or other electron beam system. As such it adds an elemental analytical capability to imaging electron beam columns. Its drawback is that the typical solid state detector has very poor energy resolution, which means that there are significant peak overlaps, sometimes making it difficult to distinguish elements. This is particularly true in the low energy region.

Focused Ion Beams (FIBs)   8

JON ORLOFF

Introduction

Focused ion beams (FIBs), in the present context, means ion beams with a focused spot size of less than 1 μm with a current density in the focused spot $J \approx 1$ A·cm^{-2}. Only in the past 15 years have such capabilities in an FIB become possible, making the FIB tool extremely useful. FIB tools are now widely used for micromachining, failure analysis, and quality assurance in the integrated circuit fabrication process, as well as for surface analysis, lithography, direct implantation, etc. In fact, in the first two of these applications, FIBs are proving to be indispensable for semiconductor device manufacturing. FIBs have recently achieved this impressive performance as a result of the harnessing of field emission ion source technology. Prior to the use of field emission technology, conventional ion sources such as duo-plasmatron sources were used, and sub-micrometer beam sizes were possible[1] only with $J \ll 1$ A·cm^{-2}. Current FIB capabilities translate to a micromachining material removal rate of ~1 μm^3·nA^{-1}·s^{-1}. This makes it feasible for FIBs to be utilized for integrated circuit dissection on a micrometer scale, for failure analysis, and for integrated circuit modification and repair.

Principles and Instrumentation

An FIB system consists of a field emission liquid metal ion source (LMIS) coupled with an electrostatic optical column, which focuses and controls the ion beam. The technology developed from efforts in the early 1970s[2–5] to develop high performance systems using gas-phase field ionization sources (GFISs). In the mid-1970s, research on LMIS began in earnest as it was realized that such sources have many very attractive properties, including the ability to operate at or near room temperature, high intensity, and very low noise.[6] LMIS were first used in an FIB system in 1978[7] and were rapidly employed for a wide variety of applications thereafter.

The heart of an FIB system is the LMIS, which generally consists of a needle with an end radius of ~10 μm coated with the liquid metal from which ions are to be produced. The needle is located in close proximity to an electrode called an extraction electrode, which creates a high electric field when a high voltage is applied to the needle relative to the electrode. The balance of forces on the liquid due to the electrostatic field stress and the surface tension force of the liquid cause the liquid to assume a conical equilibrium shape, which is called a Taylor cone.[8] Both the electrostatic force and the surface tension force are proportional to the square of the inverse of the cone radius of curvature. As the radius of the end of the liquid approaches ~5 nm, the electric field becomes high enough to cause field evaporation and ionization of the evaporated atoms. At this point a flow of liquid metal through the cone-shaped structure is established which replaces the atoms lost through ionization. Stable ion currents can be produced in the range ~1–50 μA.

LMIS are characterized by low noise, angular intensities $I' \approx 20$ mA·sr^{-1}, long life, and very stable operation. The properties required by the liquid metal to be ionized are low vapor pressure at the melting point, high surface tension, and modest chemical reactivity with the substrate needle supporting it.[9] Many metals and some non-metals have been used for LMIS operation, including Al, As, Au, B, Be, Bi, Cs, Cu, Ga, Ge, Fe, In, Li, P, Pb, Pd, Si, Sn, U, and Zn. Ga is the most popular element for micromachining, scanning ion microscopy, etc., because of its low melting point, extremely low vapor pressure, and general ease of handling. Elements that have a high vapor pressure at the melting point can be used in the form of eutectic alloys if a mass filter is available in the optical system.

Because the area at the end of the liquid cone where ionization takes place is very small, the current density is extremely large and, consequently, space charge effects are significant. The primary effects are an increase in the virtual source δ size from $\delta \approx 1$ nm for a GFIS to $\delta \approx 50$ nm for an LMIS and an increase in the fwhm of the energy distribution ΔE from $\Delta E \approx 1$ eV fwhm for a GFIS to $\Delta E \geq 5$ eV for an LMIS. In addition, ΔE increases very rapidly with the total current I drawn from an LMIS, when I is greater than ~ 2 µA,[6] while the angular intensity increases relatively slowly.

FIB instrumentation has developed along two lines. Low energy systems (E_{beam} < 35 keV) have been developed in a number of university and industrial research laboratories, whereas high energy systems (E_{beam} > 100 keV) equipped with mass filters and designed for direct implantation and lithography were developed in a few industrial laboratories. The latter instrumentation is quite expensive, and no commercial development of them is taking place at the present time (1993); rather, the emphasis is on low energy systems with pure metal sources to be used primarily for micromachining, microscopy, and surface analysis (high spatial resolution SIMS). These low energy systems have demonstrated imaging resolution of <50 nm and are routinely used for micromachining at < 100-nm beam size. Under pressure from the semiconductor device industry, many additional capabilities have been added to commercial FIBs, including SIMS, chemically enhanced etching capability, high precision specimen stages, metal deposition, light microscopy, and high resolution SEM capability.

Images are produced in an FIB instrument with secondary electrons in the same way as in a scanning electron microscope. The yield of secondary electrons is high, of the order of one electron per incident ion, and it is strongly dependent on topography and crystalline structure. The ion beam also sputters atoms from the surface being analyzed and some of those sputtered atoms are ionized. Therefore, it is possible to image the specimen with secondary ions as well as with secondary electrons.

Applications of FIBs

The main applications of FIBs are in failure analysis and circuit modification and repair in the semiconductor manufacturing industry. These applications require modest energy beams (20–30 keV), so the FIB optical columns are relatively simple. By using the micromachining capability of a submicrometer FIB it is possible to dissect a small region of a circuit to examine, for example, quality of metallization and to search for contaminants without the need to subject the device to major destructive surgery, such as chemical stripping of whole layers of material.[10] Suspicious areas in a circuit, which can be identified by analysis of device performance or by peculiar morphology, are located by means of an SEM. These areas can then be analyzed by sputtering away only cubic micrometers of material and using the FIB or an SEM to inspect the sub-surface regions. Circuits can be modified or rewired by cutting conducting runs and adding new conductors. The latter can be done by decomposing an organometallic gas with the ion beam, causing it to deposit a metallic conducting layer.[11] In order to perform these applications, it is necessary that the FIB system have an accurate and precise stage ($1:10^5$ precision) and that it be controlled by a digital computer, which directs the beam position and blanks and unblanks the beam appropriately and drives the stage.

Secondary ions are produced by the focused ion beam when material is sputtered from a surface; these secondary ions can be collected and analyzed, which is the basis for SIMS. Only a small fraction of the sputtered atoms are ionized in the process. SIMS has the advantages of great chemical selectivity and sensitivity to only the surface of a specimen. Since ions are normally produced when the focused ion beam strikes a specimen, it is natural to include SIMS as one of the capabilities of a FIB system. It is particularly useful for end-point detection and can be used, for example, when milling through a layer of passivation while seeking a conducting run in a circuit. It is also useful for producing chemical maps of specimens and for identifying small contaminating particles buried inside an electronic circuit. Since the FIB is rastered, gating techniques can be used and dynamic SIMS analysis is possible (see Technique Summary 5, Dynamic SIMS). SIMS is useful for end-point detection when milling through layered structures and for making elemental maps and analyzing small (< 1 μm) particles in semiconductor devices.

The rate of material removal with an FIB can be increased several-fold by using chemical assistance such as Cl_2 or I. When an appropriate etching chemical is sprayed on a specimen surface while machining with the FIB, it causes surface material to be removed mainly in a volatile form.[12, 13] The beam induced chemical reactions result in a much higher removal rate than can be achieved with the ion beam by itself. In addition, higher quality machined surfaces can be obtained because redeposition of sputtered material on freshly machined surfaces is minimized.

When an FIB optical column is added to a high resolution SEM, the resulting dual-beam system can be a very powerful tool for inspection. Such a system, if equipped with a field emission electron beam column, can be used for nondestructive imaging at very high resolution (<5 nm), in addition to exploiting all of the other capabilities of the FIB system. With a proper specimen handling stage it is possible to use either beam with little compromise in performance and to simultaneously perform micromachining while observing the results with the SEM. First invented in 1984,[14] dual column systems became commercially available in 1992.

Present FIB equipment costs from ~$30,000 for a stand-alone focusing column to ~$1,000,000 for the most complete system, including the various features described above. For further information on LMISs and FIBs, the reader is directed to References 15 and 16.

References

1   A. R. Hill. *Nature*, **218**, 292, 1968.

2   R. Levi-Setti. *Scanning Electron Microscopy.* **125**, 1974.

3   W. Escovitz, T. Fox, and R. Levi-Setti. *Proceedings.* Annual Meeting of the Electron Microscopy Society of America. **33**, 304, 1975.

4   J. H. Orloff and L. W. Swanson. *J. Vac. Sci. Tech.* **12**, 1209, 1975.

5   J. Orloff and L. W. Swanson. *Scanning Electron Microscopy.* **10**, 57, 1977.

6   L. W. Swanson. *Nucl. Inst. Meth.* **218**, 347, 1983.

7   R. L. Seliger, J. W. Ward, V. Wang, and R. L. Kubena. *Appl. Phys. Lett.* **34**, 310, 1979.

8   G. I. Taylor. *Proc. Royal Soc. A.* **280**, 383, 1964.

9   M. J. Bozack, L. W. Swanson, and J. Orloff. *SEM/1985/IV.* 1339, 1985.

10   R. Boylan, M. Ward, and D. Tuggle. *Proc. Int'l. Sym. for Testing and Failure Analysis.* ASM International, Materials Park, OH, 1989, p. 249.

11   A. D. Dubner, G. M. Shedd, H. Lezec, and J. Melngailis. *J. Vac. Sci. Tech.* **B5**, 1434, 1987.

12   Y. Ochiai, K. Gamo, and S. Namba. *J. Vac. Sci. Tech.* **B3**, 67, 1985.

13   A. Gandhi and J. Orloff. *J. Vac. Sci. Tech.* **B8**, 1814, 1990.

14   P. Sudraud, G. BenAssayag, and M. Bon. *Microelectron. Eng.* **6**, 583, 1987.

15   R. A. D. Mackenzie and G. D. W. Smith. *Nanotechnology.* **1**, 163, 1990.

16   J. Orloff. *Rev. Sci. Instrum.* **64**, 1105, 1993.

In Infrared Spectroscopy (IR), a beam of light (electromagnetic radiation) in the infrared wavelength region impinges on the sample and the wavelength/frequency is scanned. Whenever there is a match of the light frequency to the vibrational frequency of a vibrational mode within the sample (i.e., vibration of atomic positions) it is possible that the radiation at that frequency will be absorbed and the vibration excited (selection rules apply and not all vibrations can be excited). The absorption occurring, as a function of IR wavelength, is monitored by comparing the input intensity of the radiation to the output intensity, thus revealing the vibrational frequencies existing in the sample. Since a vibrational frequency value is dependant on the bonding between the atoms involved in that vibration, vibrational frequencies of materials are characteristic of the chemical groups existing in the material (e.g., OH, CN, CH_2, COOH, etc.), so determination of the vibrational frequencies present allows determination of which chemical groups are present. With standards the concentrations of the groups present can also be determined from the strength of the absorptions.

Fourier Transform IR simply describes the most common modern technique for scanning the wavelength of the impinging IR beam. It is based on the Michelson interferometer, where constructive and destructive interferences between the two halves of a split light beam are controlled by changing the path length of one of the beams with respect to the other by moving reflecting mirrors. Since the path length difference for destructive/constructive interference depends on wavelength, the oscillation of the mirrors provides a way of scanning the wavelengths from a polychromatic light source.

For solids, IR spectroscopy can be performed either in reflection or in transmission if the sample is thin enough to pass the radiation. This varies enormously with material because absorption coefficients vary enormously (some materials are transparent to IR, some highly absorbing). In reflection, coupling the radiation to an optical microscope allows a lateral resolution down to 20 μm, with the depth probed depending on absorption coefficient, but usually being many μm. Another reflection mode is Attenuated Total Reflection, ATR, where light enters at grazing angle and probes only the surface region (10's of nm). IR of solids does not require a vacuum, but the path length of the radiation through ambient atmosphere must be purged using dry nitrogen, to avoid gas phase absorption from strong absorbers such as water vapor.

The strength of IR is the qualitative and sometimes quantitative (with standards) identification of the presence of chemical species, or functional groups, sometimes down to trace levels in liquids and solids. For solids stress, strain, crystallinity/amorphousity, and inhomogeneities can be detected from absorption peak broadenings and shifts.

Hall Effect Resistivity Measurements 10

GEORGE N. MARACAS

Resistivity and carrier concentration and mobility are easily obtained by Hall effect measurements. Temperature-dependent Hall measurements provide carrier compensation ratio, total impurity density, ionization energy, and dominant scattering mechanism. Extending the technique to high variable magnetic fields (~12 T) allows magnetoresistance characterization of mobile carriers in two-dimensional conducting channels and of quantized state energies.

Samples require a well-defined conducting-layer thickness or conducting channel. Isolation of the channel from the substrate can be achieved by ion implantation or epitaxial growth. No isolation is required if bulk material (e.g., substrates or thick slabs) is to be characterized. Ohmic contacts are placed onto patterns such as a square slab, Van der Pauw (VDP or clover-leaf) pattern, or Hall bar. The square slab is the easiest, the VDP has the best contact placement error immunity, and the Hall bar is the most appropriate for magnetoresistance measurements. Sample dimensions are on the order of 1 cm.

The Hall effect technique uses a four point probe geometry and exploits the Lorentz force behavior of mobile carriers in an applied magnetic field to extract the "Hall coefficient," R_H, which is directly related to the carrier density. By combining concurrent resistivity measurements on the same sample, extraction of free carrier concentration, type, and mobility are possible. In the simplest method, a constant current is applied between two contacts and the voltage is measured across the other two. Application of a magnetic field (~2 kG) creates a charge density imbalance that produces a voltage across the two contacts located perpendicular to the current flow direction. The magnitude of the voltage is directly proportional to the mobility; the polarity indicates the carrier type.

Mobility depth profiling can be accomplished by either including a rectifying "gate" or by chemical etching, which controls the conducting layer thickness between the contacts. Numerical techniques are required to extract the uniform or non-uniform doping/mobility profiles versus depth.

Commercial Hall effect systems automatically control the multiple contact terminal switching sequence and magnetic field application. The cost of an electromagnet raises the price of systems; some are available with small fixed magnets. Electrometers or very high impedance buffers are required to measure semi-insulating materials, which also raises the price of a system. Temperature-controlled cryostats (4 K to above room temperature) are also available and vary considerably in design and cost.

Information	Carrier concentration, mobility, type
Element range	Not element specific

Destructive	Yes
Depth profiling	Yes, limited
Depth resolution	Limited by Debye length
Depth limit	Barrier breakdown voltage depth; uniform etch depth limit
Doping range	$\sim 10^{14}$ to high 10^{19} cm^{-3}
Sample requirements	Four contacts, ~ 1 cm sample size
Cost	\$10,000–\$50,000 (depends on magnet size and type and buffer impedance); high field magnets: $\sim$\$35,000

Recommended Reading

Blood, P., and J. W. Orton. *The Electrical Characterization of Semiconductors: Majority Carriers and Electron States*. Academic Press, New York, 1992, Chapt. 4.

In Inductively Coupled Plasma Mass Spectrometry (ICPMS), the sample is introduced by one of several ways into an inductively coupled plasma which fragments and ionizes all species present down to the atomic level. The ions are passed into a mass spectrometer (several types are in use) and mass analyzed to reveal which atoms are present. The plasma is highly efficient, and, coupled with an efficient and high resolution mass spectrometer, detection limits down to sub ppb are achievable for many elements. If a low resolution mass spectrometer is used, mass interferences limit the detection limit for some elements, notably Fe. Rough quantification to the 5 to 20% level is easy using standard sensitivity factors determined for the mass spectrometer. For better quantification standard solutions using a material matrix as similar to the sample as possible are used. Sub 1% accuracy can be achieved by using the method of isotope spiking.

The normal method of sample introduction is to nebulize an aqueous solution into the plasma. For solids, dissolution by a solvent, or digestion by an acid (normally nitric acid), is the preferred method, but direct sampling using laser ablation is possible. This method is the only way to achieve any spatial resolution. Laser ablated craters are typically 10's of μm wide and a fraction to a few μm deep.

The main uses of ICPMS are in two areas. One is the routine determination of contaminant levels in solutions and solvent, particularly water supplies. The second is for determination of trace levels of elements in semiconductors and alloys.

Light Microscopy* **12**

The visible light microscope dates back over 300 years for the study of natural materials (plants, animals). At its simplest level the method involves visible observation using white light, with useful magnification of up to 1400X. This allows observation resolving features down to around 0.2 μm, as opposed to about 200 μm by the naked eye. Resolving power can be increased by using UV radiation (doubled) or by immersing optics in oil, but both procedures greatly increase the compexity of instrumentation. For biomedical samples and many other materials (e.g., fibres, wood, hairs, pollens), the morphology revealed in the image is the main information, and materials identification can usually be made through use of extensive atlases of every conceivable structure observed. For materials science, simple morphology alone is often insufficient for identification, and it is necessary to investigate a number of other parameters such as opacity, color, refractive index, crystal system, fluorescence, and many others. These can be quickly determined using a variety of procedures and attachments to a microscope or several microscopes.

The main role of the light microscope is to image structures, either natural or man made, which are not observable to the naked eye, but have features greater than 0.1 μm. It is extensively used during the preparation of samples for observation by other analytical techniques, such as SEM or AFM, and should always be the first step in a materials analysis.

Low-Energy Electron Diffraction (LEED) is a technique for providing a two-dimensional diffraction pattern from a surface. It is only applicable to a surface with long-range order, typically a single crystal surface, or an adsorbed over layer. The method involves impacting a collimated beam (typically 0.1 mm diameter) of monoenergetic electrons, in the energy range 10 to 1000 ev (equivalent to a photon wavelength of 0.4 to 4 Å) onto the surface at normal incidence. Electrons in this energy range have very short inelastic mean free path lengths, so the electrons which are elastically diffracted back from the sample only come from the first few Å depth. They are separated from inelastically scattered electrons, which have lost energy, by a retarding grid system and are then detected on a phosphor screen. The observed diffraction pattern reveals the 2D unit cell type and dimensions. Determination of the actual surface atomic atom positions relative to each other, or the underlying layers, requires measurement of the diffraction intensities as a function of varying the electron energy and comparison to theoretical calculations on trial structures. The shapes, splittings, and broadening of the diffracted beams also carries information on surface disorder.

The major use of LEED has been to determine surface structures of both clean bulk crystalline material and adsorbed layers on such surfaces, by comparison to theory. It is also used routinely and simply to monitor the surface cleanliness of single crystal surfaces, and how well ordered the surface is (sensitive over about a 200 Å range in most instruments). A very specialized version exists, the Low Energy Electron Microscope, LEEM, which functions just like an imaging TEM and has a surface imaging resolution of 150 Å.

Neutron Activation Analysis (NAA)*   14

Neutron Activation Analysis (NAA) is a technique for trace analysis in bulk material. It involves two separate steps. The sample is initially irradiated by neutrons in a nuclear reactor to create the radio isotopes (which may take 1 to 7 days). This is followed by removal to a gamma-ray spectroscopy facility, where the gamma-rays given off during subsequent decay of the radio isotopes can be monitored. The gamma-ray energies emitted are characteristic of the atoms decaying and are determined using a solid state energy dispersive detector in an analogous manner to the X-ray analysis in EDS (see appendix 7), except that the energies are in the MeV range. The technique is a bulk one, since both neutrons and gamma-rays penetrate deeply. There is also no spatial resolution.

Both the initial neutron capture probability (cross-section) to create the radio isotopes, and the half-lives for the subsequent decay process vary enormously, in no particular pattern, across the periodic table. In fact, only about two thirds of the elements can produce radio isotopes this way. So trace element sensitivity varies enormously among elements, and also with the matrix, which may give off competing gamma-rays. For trace metals in Si or SiO_2, ppb to ppt sensitivity is easily obtained, but other semiconductor matrices such as GaAs are less favorable. Trace elements in plastics and biological materials can be measured because the matrix elements of C, O, H and N all have low neutron capture cross-sections. The material may suffer radiation and heating damage in the process, however.

The major uses of NAA are for trace determination of bulk impurities in semiconductor, biological, geological, and envonmental samples, and in forensic science. Quantification can be achieved from signal intensities using standards, or from calculations if enough information about the instrumetal conditions is available. The major drawbacks are the need to use reactor and certified gamma-ray facilities, the long time involved for an analysis, and the special handling needed for the samples.

Optical Scatterometry* **15**

Optical Scatterometry determines the angular-resolved distribution of light (usually a visible wavelength laser) scattered from a surface. The distribution is a function of the properties of the bulk material (refractive index, homogeneity, orientation, density), the surface roughness, the surface contamination, and the source polarization and angle of incidence. The ratio of the scattered intensity to incident intensity as a function of scatter angle is the parameter determined.

On nominally smooth surfaces, measurement at a specific angle corresponds to determining the root mean average roughness at a particular spatial frequency (the intensity of scattered light is proportional to the roughness amplitude). Thus both long range and short range roughness can be determined, down to the A level, and with a spatial resolution of half the wavelength of the laser light.

Periodic structures (e.g., lithiographic etched lines) may also be characterized. Light is scattered into different diffraction orders, with the intensities in the different orders being strongly affected by the shape of the periodic struture (e.g., depth of etch, slope of sidewall).

The major use of Scatterometry is to characterize the surface roughness of nominally smooth surfaces and as a process metrology tool for lithography in wafer processing. It is fast, non-destructive, and can yield quantitative results when coupled to theoretical modelling. The disadvantage is that it gives an average picture and cannot observe any individual feature, unlike techniques such as SEM or AFM.

In Photoluminescence (PL), light in the visible wavelength region (usually laser light) is used to induce excited electronic states in the material concerned, and the light intensity then re-emitted as these states relax is monitored, usually as a function of wavelength, by passing through a lens and an optical spectrometer onto a photodetector. Another option is just to collect the total light emmitted, and a third option is to vary the wavelength of the light doing the excitataion.

The energy (wavelength) of a photoluminescence peak in the spectrum of intensity versus wavelength is determined by the difference between the enery of the excited state and the ground state to which it relaxes. Unlike X-ray fluorescence, where the energy levels involved are atomic in nature, in photoluminescence the energy levels involved are, for solids, the conduction bands (excited states) and valence bands (ground states), which are characteristic of the bonding effects in the material but not the individual atoms. No direct atomic identification is given, but peak energies relate to such properties as band gap, chemical compound, defect and impurity electronic states present, disorder and strain. It is usually qualitative rather than quantitative, though by comparison to standards it is possible to quantify trace impurity states. The depth probed varies from about 0.1 µm to several µm, depending on the light absorbtion properties of the material. Using a microscope, spatial resolutions down to the 1–2 µm range are possibe.

Major practical uses of PL are for determining band gaps in films and general defect detection. Impurity levels down to ppt levels can sometimes be detected. The mapping of the PL of wafers can be performed as a quality control method, particularly with respect to defects.

In Raman spectroscopy, a light beam of fixed wavelength (usually in the visible range) from a laser undergoes inelastic scattering as it interacts with the sample material. The photons can suffer energy loss if they excite vibrations in the sample. The energy losses, and therefore the vibrational frequencies involved, are determined by measuring the wavelengths of the scattered light. The value of an excited frequency is sensitive to the bonding between the atoms involved in that particular vibration.

Raman spectroscopy is closely related to infrared spectroscopy, which also determines the values of vibrational frequencies (see the FTIR summary), but it differs in the physics in that it involves a photon scattering process with energy loss, where as IR involves varying the photon energy and observing absorption occurring at the vibrational frequencies. Raman spectroscopy is better suited to optical microscopes, because of the single laser wavelength involved, and spatial resolution down to the one µm range is possible (cf. approximately a 10 µm limit for IR).

The depth probed depends very strongly on the optical properties of the material involved and the laser wavelength used. Moving into the UV range greatly reduces the probing depths, and allows a depth profiling capabilities if several wavelengths are used.

The major area of application for solids and liquids is chemical fingerprinting and the identification of unknown compounds. For solids, Raman is also used for phase identification, following amorphous/crystalline transitions, measurement of stress and strain, and, in the microscope mode, the detection and analysis of defects, including particles during wafer processing.

Reflection High-Energy Electron Diffraction (RHEED)*

In Reflection High Energy Electron Diffraction (RHEED), a beam of electrons in the 5 to 50 KeV energy range strikes the surface of a flat solid sample at grazing angle (1–5 degrees). The wavelength of these electrons is shorter than interatomic distances, and so diffraction of the beam takes place if there are ordered rows of atoms present at the surface of the solid. The diffracted beams are scattered into directions which depend on the spacing between rows (cf. LEED or X-ray diffraction), and the subsequent diffraction pattern is observed on a phosphor screen. The symmetry in the RHEED pattern reveals the symmetry of the atomic rows causing the diffraction, and the spacings in the pattern reveal the lattice spacings between rows. Broadenings and splittings of the diffraction features can also provide information on the degree of lattice disorder at the surface.

Owing to the grazing incidence, only the first few atomic layers of the surface are probed, despite the relatively high electron energies involved. Owing to the requirement for grazing incidents, very flat surfaces are needed (e.g., Si wafers), and there is very limited lateral resolution (typically 0.5mm by 4mm beam area on the surface).

The main use of RHEED is to monitor surface crystal structure during thin film epitaxial growth on single crystal surfaces. It is very sensitive to defects, and can distinguish 2D defects from 3D island growth.

In Rutherford Backscattering Spectrometry (RBS), a solid is bombarded with a monoenergetic beam of ions (typically He or H) in the one to several MeV energy range. As the beam penetrates into the solid, the ions undergo continuous small angle forward scattering through interactions with the electron density between the atomic nuclei, continuously losing small amounts of energy as a function of depth penetrated. In addition, a small fraction of the ion beam undergoes head on collision with the nuclei of the atoms present in the solid (the target atoms) and is consequently back scattered. The back scattering results in a major loss of energy (transferred to the atoms struck, which moves forward like in billiard ball collisions) from the probe ion beam, the amount of which is characteristic of the mass of the target atoms. Determination of the energy losses suffered by the back scattered ion beam after exit from the solid thus allows, in principle, identification of the atoms present, their depth distribution, and, from intensities, the atomic concentrations. The depth probed can vary from 1–20 µm depending on ion species used, ion energy, and instrumental set-up. At these high energies, the ion/material interactions are well described by classical kinematic equations, and quantification can be achieved without standards. The depth resolution also varies strongly, depending on the material, ion energy, and instrumental set-up, plus the depth probed (poorer at greater depth) ranging from a best of about 2 nm to 30 nm. Element sensitivity is a strong function of the position in the periodic table, varying as Z squared. It can be as poor as 10 atomic percent at the low Z end and as good as 10 ppm range at very high Z. In principle all Z values above that of the probe ion can be detected (so not H), but the poor energy resolution of the solid state detector used means that some elements cannot be distinguished from adjacent elements in the periodic table. Lateral resolution is limited to the beam diameter, usually 1 to 4 mm, but a few specialized microbeam systems do exist.

The main use of RBS is nondestructive element depth profiling of thin films in the few µm depth range. If the target is crystalline, specialized modes involving beam alignment with crystallographic axes can give structural information, such as the degree of lattice damage.

In Scanning Electron Microscopy (SEM), a probe electron beam (typically a few hundred eV to a few 10's keV energy) is finely focused (down to 10 Å capability in some instruments) and scanned over a solid surface. The interaction of the beam with the sample material generates a variety of responses, including fluorescence emission, which can be used for elemental analysis (see the EDS summary).

One of the major responses is a copious emission of low energy secondary electrons (0 to 50 eV range), which escape from the top few 10's Å of the material. The secondary electron yield strongly depends on the angle of impact between the probe beam and the local surface topography, so rastering the beam across the surface produces a changing intensity with changing topography. These changes in secondary electron emission intensity are used to modulate the brightness of a synchronously rastered cathode ray tube, creating an image. The image can be highly magnified up to 500KX, with a lateral resolution determined by the diameter of the probe beam, and have the look of an optical image (though the depth of focus is much greater).

In addition to the secondary electrons, a much smaller quantity of backscattered primary electrons are produced. Their intensity depends strongly on material (high for high Z elements), so the image also contains some Z contrast. Filtering out the low energy secondaries enhances this contrast (backscatter mode), producing an average Z dependant image instead of topography image.

The SEM is often the first or second (after an optical microscope) technique used to provide a magnified image of an area of the sample to be examined. It is often used in conjunction with an ancillary analytical technique, such as EDS, to provide elemental analysis capability to go with the imaging.

Scanning Transmission Microscopy (STEM) is a specialized form of TEM (see TEM summary) where the high energy electron beam (100 to 300 keV) is focused to a small spot (down to a minimum of about 3 Å) and scanned across a sample which is usually a material which has been sectioned to a very thin film (5 to 500 nm thick). After interaction with the material of the film, the transmitted beam can be subjected to various treatments, as in TEM (bright and dark field imaging; convergent beam electron diffraction, CBED; and electron energy loss spectroscopy, EELS). Also, the emitted X-ray fluorescence can be analyzed by EDS (see EDS summary). Keeping the sample to a very thin section stops the lateral spread of the primary beam as it passes through, allowing the spatial resolution to be determined by the diameter of the focus beam, down to almost atomic dimensions.

The major use of STEM is for imaging and compositional analysis (sometimes also chemical state analysis, from EELS) at very high spatial resolution across thin film material interfaces, such as found in semi-conductor and other high technology devices. Single atom imaging is possible for heavy atoms. The disadvantage is that a very thin interface cross-section must be first prepared. Also, at very high resolution, the electron beam is focused into such a small area that the local electron density becomes extremely high and can anneal, damage, or even burn through the film if the dwell time is too long.

In Scanning Tunneling Microscopy (STM), a sharp tip is brought to within a few Å of a conducting surface in ultra high vacuum and held at a positive or negative potential with respect to the tip. At this distance a quantum mechanical electron tunneling current flows between an individual atom on the surface and an atom on the tip (or vice versa, depending on the polarity). The magnitude of the current is an exponential function between the two atoms involved. So scanning the tip across the surface therefore produces a strongly varying current as the topography and distance between tip and surface atoms changes. This allows mapping of the topography (actually the surface electron density) with a resolution of 0.01 Å in the Z direction, and atomic resolution in the XY direction.

In the related technique, Scanning Force Microscopy (SFM; also known as atomic force microscopy, AFM), the tip is mounted on a cantilever and the Van De Waals forces between the surface and the tip deflects the cantilever. The deflection is also a strong function of the separating distance, allowing topography mapping, as in STM. The resolution in SFM is about ten fold poorer than STM, but the technique has the great practical advantages of being operated in air and not requiring a conducting sample.

The major use of STM is in the research area of imaging atomic structures on clean conducting and semi-conducting surfaces, under UHV conditions. SFM, on the other hand, is widely used throughout industry as a very high resolution surface profilometer to monitor surface roughness, defects, and man-made micro or nano structures.

WALTER JOHNSON and CHUCK YARLING

Sheet resistance measurements are a way of electrically characterizing semiconductor wafers, thin films on wafers, and thin near-surface regions of diffused or ion-implantation doped wafers. The method of choice for the resistance measurement is a four point probe arrangement of contacts with the surface through which current is passed using two of the probes and the voltage drop being measured across the other two.

Parameter measured	Electrical resistivity
Main usage	Quality control of Si wafers at various processing stages
Sample requirements	Flat surface, a few square millimeters in area
Destructive	No, unless depth profiling by beveling sample
Accuracy	1%
Quantitative	Yes, but calibration needed
Lateral resolution	mm usually
Depth resolution	None, except by beveling sample
Instrument cost	$10,000–$100,000
Size	Benchtop

Introduction

The four point probe is the most common tool used to measure sheet resistance. The normal range for sheet resistance in semiconductor processing is from less than 0.02 Ω/square for aluminum films to about 1 MΩ/square for low dose implants into silicon.

Until the advent of the dual configuration technique,[1] the construction of a high precision tester,[38] and control over contact resistance, accurate sheet resistance measurements were usually only possible with van der Pauw structures. These advances brought the consistency to a level where the four point probe could be used for equipment and materials characterization. Sheet resistance is a widely used method for measuring the electrical properties of films in the semiconductor industry. Typical applications include starting silicon substrates, ion implantation, diffusion, epitaxial silicon, polycrystalline silicon, metals and silicides.

Principles and Definitions

Materials can be classified by their ability to conduct electricity[3]: insulator, conductor, and semiconductor. Insulators cannot be measured with the four point probe and often are the source of problems in measuring conductive or semi-conductive films.

Metals tend to give up valence electrons freely for conduction, resulting in all the electrons bonding all the atoms in an electron sea.[4] Electrical resistivity results from scattering of these electrons by the lattice.[5]

Silicon is an insulator at a temperature of absolute zero (0 K). There are no free carriers to conduct current. As temperature rises within the bulk material, bonds within the silicon lattice structure begin to break up, providing the free carriers necessary for current flow. Both n, the number of electrons per cubic centimeter, and p, the number of holes per cubic centimeter, depend on the temperature of the material. Since each broken bond produces both an electron and a hole, both n and p are present in equal numbers in pure silicon. Thus, at room temperature (T = 27 °C = 300 K),

$$p = n = n_i = 1.45 \times 10^{10}/\text{cm}^3$$

Donors and Acceptors

As previously stated, intrinsic silicon has an equal number of donors, n, and acceptors, p, the number of which is dependent upon the ambient temperature. However, it is the interaction of adjacent semiconducting layers having different densities of dopants that give semiconductor materials their unique properties. These adjacent layers are usually, but not necessarily, doped with carriers of opposite dopant types. Doping may take place by several methods, including diffusion, implantation, chemical-vapor deposition (CVD), and spin-on glass (SOG).

Dopants are impurity atoms added to the semiconductor silicon during crystal growth or subsequent processing to enhance its ability to carry current (its "conductivity").[3, 6] For example, silicon is doped with arsenic or phosphorous (and sometimes antimony) to produce an n-type material. For p-type materials, silicon is normally doped with boron. The total number of donor, or n-type, dopant atoms is N_d. The total number of acceptor, or p-type, dopant atoms added to silicon is Na

Mobility

Mobility, or drift mobility, μ, is strongly dependent upon process conditions (such as anneal parameters) for high dopant conditions.[3, 6] Mobility is defined as the "ease" with which the carriers move through bulk material. Conditions within the bulk material that may affect this property include carrier concentration, temperature, crystal defects, presence of undesired impurities, and other scattering mechanisms.

Mobility is quantitatively defined by the following formula:

$$\mu = D \cdot (q/kT), \qquad \text{cm}^2/\text{V} \cdot \text{s}$$

where D is the diffusion constant, a bulk property of silicon, in cm^2/s; q is the electronic charge; k is Boltzmann's constant (1.38×10^{-23} J/K); and T is temperature in kelvin.

When the dopant type and total impurity concentration of a silicon sample are known, the mobility and diffusion coefficient may be obtained from calibration curves relating mobility to dopant concentration.[3, 6]

Conductivity

Once impurities have been added to the silicon, the material is now "quasi-" or "semi-" conducting. Conductivity, σ, is the ability of the silicon to conduct electron and hole flow.[3, 6] Conductivity is defined:

$$\sigma = (qn\mu_n) + (qp\mu_p), \quad (\Omega \cdot cm)^{-1}$$

where q is the charge of an electron (1.6×10^{-19} coulombs); n is the number of electrons in carriers/cm^3; p is the number of holes in carriers/cm^3; μ_n is the electron mobility in cm^2/V·s; and μ_p is the hole mobility in cm2/V·s.

In most cases, one of the dopant types, either donor (N_D) or acceptor (N_A), far exceeds the number of the other. When this happens, this equation is reduced to only one term. The equation then becomes

$$\sigma = (qn\mu_n), \quad (\Omega \cdot cm)^{-1} \quad \text{if } N_D \gg N_A$$

or

$$\sigma = (qp\mu_p), \quad (\Omega \cdot cm)^{-1} \quad \text{if } N_A \gg N_D$$

Resistivity

The resistivity of a material [7, 8], is a measure of its ability to oppose electrical conduction that is induced by an electrical field placed across its boundaries. Resistivity, p, is defined as

$$\rho = 1/\sigma = \sigma^{-1}, \quad \Omega \cdot cm$$

The resistivity of the three classifications of materials used in semiconductors is as follows:

insulator	$\rho > 10^3$ $\Omega \cdot cm$
semiconductor	$10^{-3} < \rho < 10^3$ $\Omega \cdot cm$
conductor	$\rho < 10^{-3}$ $\Omega \cdot cm$

Resistivity also varies with temperature. Temperature coefficients of resistance (TCR) can be found in the literature.[9] Pure metals commonly used in semiconductor manufacturing have TCRs in the range of +0.2 to +0.4%/°C.[10] There are several cases where negative temperature coefficients of resistance can be obtained for very thin, low density or cermet films.[5] Doped polysilicon layers can also exhibit negative TCRs.

Resistance

Once the resistivity of a silicon sample is known, its resistance may be determined. Resistance, R, is defined as

$$R = \frac{\rho L}{A} = \frac{\rho L}{tW}, \quad \Omega$$

where ρ is the resistivity in $\Omega \cdot$cm; L is the length of the sample in cm; and A is the cross-sectional area of the sample being probed in cm^2 = thickness (cm) $\times$ width (cm) = t (cm) $\times$ W(cm).

Sheet Resistance

Free-standing thin layers can be usefully described with the term sheet resistance, R_s, rather than by the bulk resistance. This is more descriptive, since it involves the thickness, t, of a layer. Sheet resistance[11] is calculated from the resistance of the sample by using the formula above, which for $W = L$ reduces to

$$R_s = \frac{\rho \ (\Omega \cdot \text{cm})}{t \ (\text{cm})} \quad \text{in } \Omega, \text{ but more commonly referred to as } \Omega/\square$$

The value of the concept of sheet resistance lies in the ability to calculate the resistance of large areas knowing only the size of the sample and its resistivity. For instance, if a bar of silicon is twice the length of another bar (of the same resistivity), the resistance of the longer bar is twice that of the shorter one, but the sheet resistance, R_s, remains the same.

The Four Point Probe Method of Measuring Sheet Resistance

Probe assemblies for the four point probe historically have probe tips which force a current and measure the resulting voltage drop, for the sheet resistance measurement, in one of two arrangements: linear or square arrays. Since the linear probe-tip array that is perpendicular to the edge of the wafer is the most prevalent, only this arrangement will be considered in this discussion.

Sheet resistance is measured by passing a current between two probe pins and measuring the resultant voltage drop across two other pins. In the linear array, it is customary for the outer pins to carry the current and the inner two to measure the voltage. Geometric effects that interfere with the sheet resistance measurement, such as the probe's proximity to the edge of the wafer (which causes "current crowding"), can be accounted for by a technique called dual configuration. This technique requires that two resistance measurements be made.

The first measurement, R_A, is obtained in the traditional manner, as shown in Figure 1a, with the outer pins carrying the current and the inner pins measuring the voltage. The second measurement, R_B, is obtained by having the first and third pins carry the current and using pins two and four to measure the voltage, as shown

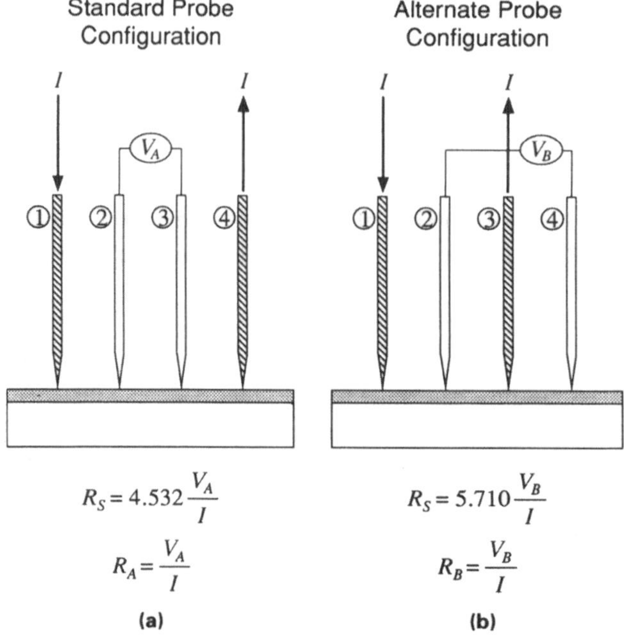

Figure 1   **Paired measurements which are used to calculate sheet resistance in the dual configuration mode.[1]**

in Figure 1*b*. The use of both arrangements allows for the correction of discrepancies in pin spacing and any current crowding effects.[1]

Measurement Considerations

Selecting the optimum probe head and measurement current for some samples can be a difficult task. Usually we settle for some acceptable set of conditions that are determined with the minimum of effort. Metal films seldom exhibit measurement problems due to the good ohmic contact created in the metal pin to metal film contact. For some samples we need to exercise a little more care in investigating the conditions on the wafer and the type of probe used to find conditions that give the most accurate and repeatable measurements. For example, in the case of an implanted layer, a few things to consider are dose, energy, species, background concentration, screen oxide, and surface preparation. There are a number of combinations that create conflicting measurement requirements, thereby necessitating compromise.

Requirements for an ideal measurement:
- minimum contact resistance
- maximum surface concentration
- minimum leakage across the surface
- minimum leakage across any layer boundary or junction.

If all the above requirements are met, then there is a good chance of getting excellent results. Some ways to achieve these requirements are discussed below.

Minimum contact resistance is obtained by ensuring that any unwanted insulator layer, such as a thick oxide, is removed from the surface; deposited oxides will cause measurement problems if not effectively removed prior to measurement. Native oxides will be present on most layers, but in most cases the standard 100 mg loading probe head will effectively puncture this native oxide to make good contact at the conductive layer. This does rely, however, on the probe head having some surface roughness or on its being sufficiently sharp. An inversion layer common on p-type silicon substrates[12, 13] usually can be bypassed by using a sharp probe with high gram-loading that can break through this layer to make contact with the more conductive layer, the one we wish to characterize. More current can sometimes help break down this layer.

High contact resistance can result from contaminated probes, particularly if the probes are used on metallic layers and the metal has transferred to the probe tip.

Another cause of high contact resistance is low surface concentration of dopant in diffused or implanted layers. This effect will normally manifest itself in poor probe qualifications; it can also give rise to random bad site measurements or nonreproducible maps. If the high-resistance surface is part of the layer measured, then increasing the surface area of the four-point probe pins that contact the sample will decrease the total contact resistance (i.e., use a larger probe tip radius).

Maximum surface concentration can be obtained for implant monitors by reducing the background concentration (i.e., using a higher resistivity substrate). This increases the resultant concentration after the implant. A screen oxide can also be used, capturing the low-concentration tail into the oxide layer. This tail is consequently removed upon removal of the oxide.[1]

If low surface concentration seems to be the problem, the first requirement is to better understand the physical characteristics of the layer being measured. Parameters such as species, dose, energy, background concentration, screen oxide, and surface preparation should be considered, as well as the use of simulations predicting the likelihood of success. In some cases sharper probes or higher loadings may be recommended; in other cases, different surface treatments may be tried or alternative substrates recommended for the test wafers. In all cases, the objective is to be sure to characterize the material over the depth of interest, not a surface region which may vary but is of no device performance interest (e.g., native oxide or a substrate material).

Native oxides incorporate a number of various contaminants and can often cause poor sheet resistance measurements, especially on silicon where the resistivity is relatively high. The surface leakage is reduced by ensuring a clean, fresh surface with a very thin, stable SiO_2 layer on silicon. Typically this involves an HF dip, rinse, H_2SO_4–H_2O_2, and rinse procedure for silicon wafers.

The maximum substrate junction resistance would, of course, be obtained by putting the conductive layer on an insulator. This is not always possible or practical. For implants, the only real alternative is to use a monitor wafer of opposite conductivity type with a lower background concentration. An epi junction width can be narrowed with a high resistivity substrate or a process change. High resistivity substrates will help, where metal layers must be deposited directly on silicon.

Troubleshooting

Some problems commonly encountered making sheet resistance measurements are:
- oxidizing metal layers
- punch-through of thin layers by the probes to other layers or the substrate
- wear of probe tips, making them smoother and rounder
- contamination of probe tips, increasing their contact resistance
- measurement current too high, causing layer heating and an apparent rise in sheet resistance
- measurement current too high, causing substrate leakage and an apparent fall in sheet resistance
- substrate resistivity too low for the implanted dose, causing substrate leakage.

A good starting point in any troubleshooting procedure is to refer to a known standard. In this case, it is desirable to set aside test wafers (or "golden wafers") which have undergone the same process as those being currently tested and which have given good measurement results in the past. Re-measurement of the test wafer will now give an immediate sign of whether the problem lies in the wafer or the probes. Although troubleshooting is possible without such wafers, the task becomes more time-consuming.

When trying to make measurements on shallow implants and diffusions or thin metal layers, it may be possible to "punch-through" the layer when the probes used are too sharp. This allows leakage into the substrate, effectively providing current paths in parallel and giving apparently lower sheet resistances. The remedy is to use a flatter probe or higher resistivity substrates. In some circumstances we need to compromise on a set of conditions that gives, as before, reproducible measurements, as shown by good probe qualifications.

After extended use, probe tips become worn, and even after probe conditioning, the surface irregularities that break through the native oxide will no longer be present. It will then be difficult or impossible to get acceptable results. When this occurs, the probe should be replaced.

Probe tips will inevitably pick up contamination after use, either from the wafers they probe, or from handling, or from the air. The usual result of probe contamination is an increase in the number of bad sites during mapping and poor probe qualifications. To remove lightly bound contamination, brush the tips lightly with a fine brush or a Q-tip soaked in a fab solvent such as IPA. As usual, a probe qualification will reveal the effectiveness of this treatment. Tightly bound contamination may require a probe conditioning, the necessity of which will vary with each application.

Probe Qualification

One of the main sources of noise in the sheet resistance measurement is the probe itself. To get accurate and repeatable results, the probe head pins must make good electrical contact each time they touch the surface of the wafer. The probe qualification procedure should check the short-term repeatability of the probe head.

For example, if a probe used to measure a process with a uniformity specification of 0.75% has a repeatability (noise) of 0.5% and the true wafer uniformity is 0.65%, then the net result is

$$S = (S_{\text{wfr}}^2 + S_{\text{probe}}^2)^{1/2}$$

where S is the resulting standard deviation, S_{wfr} is the real standard deviation of the wafer, and S_{probe} is the probe qualification standard deviation.

The monitor standard deviation of 0.82% is larger than the uniformity specification and therefore out of specification. From this example we may conclude that the tighter the process monitor specification, the smaller should be the allowable noise introduced by the probe head. As the probe repeatability degrades, the quality of the map degrades first. The second parameter to degrade is the standard deviation; the last is the average sheet resistance. To assure the continued quality of the measurements and maps, the probe repeatability should be checked regularly. It is also essential to use the right probe head for each application. Probe repeatability is best checked on monitor wafers specially chosen for this purpose. If the repeatability is being checked on ion implanted wafers, a wafer should be used that has a resistivity equal to or greater than that of the wafers typically measured. Depending on the process monitored, it is essential to keep a few monitor wafers set aside. These monitor wafers should be changed about every three months, depending on their type and usage and on the type of probe head used.

Conclusion

Many factors govern the success or failure of sheet resistance measurements. Methodology exists for troubleshooting poor or doubtful sheet resistance results. It is useful to consider the following generalized statements when preparing to measure a wafer or investigating problems:

- When investigating suspect results, perform a probe qualification or repeatability on the suspect wafer. If the probe qualification is good, then the results are believable. If the probe qualification is bad, carry out a second probe qualification on a monitor wafer with similar process history and sheet resistance. If these results are good, then the wafer has a problem; if bad, then the probe has a problem.
- A good probe qualification usually identifies a probe that will give a meaningful map with explainable features and accurate sheet resistance.
- A map with inexplicable rejects or erratic high and low points usually means a measurement problem and erroneous standard deviation.
- If there is a contact resistance problem, a poor probe qualification will result.

This summary relates closely to Technique Summary 24, Spreading Resistance Analysis (SRA). SRA refers specifically to a resistivity, or sheet resistance, profile made along a beveled surface cut from the sample; that is, it measures resistivity as a function of sample depth. The use of two point probes for making this measurement is described in the SRA summary. Four point probes can also be used and, for flat surfaces, is the preferred method.

References

1. D. S. Perloff, J. N. Gan, and F. E. Wahl. "Dose Accuracy and Doping Uniformity of Ion Implantation Equipment." *Solid State Tech.* **24** (2), 1981.

2. W. A. Keenan, W. H. Johnson, and A. K. Smith. "Advances in Sheet Resistance Measurements for Ion Implant Monitoring." *Solid State Tech.* **28** (6), 1985.

3. P. E. Gise and R. Blanchard. *Semiconductor and Integrated Circuit Fabrication Techniques.* Reston Publishing Co., Reston, VA, 1979.

4. M. J. Starfield and A. M. Shrager. *Introductory Materials Science.* McGraw-Hill, New York, 1972.

5. L. I. Maissel and R. Glang. *Handbook of Thin Film Technology.* McGraw-Hill, 1983.

6. R. S. Muller and T. I. Kamins. *Device Electronics for Integrated Circuits.* Wiley, New York, 1977.

7. J. C. Irvin. "Resistivity of Bulk Silicon and of Diffused Layers in Silicon." *Bell System Technical Journal.* **41** (2), 387–401, 1962.

8. E. Hesse. "Resistivity Measurements of Thin Doped Semiconductor Layers by Means of Four-Point Contacts Arbitrarily Spaced on a Circumference of Arbitrary Radius." *Solid State Electronics.* **21**, 637, 1978.

9. "Standard Method for Measuring Resistivity of Silicon Slices with a Collinear Four-Probe Array." *Annual Book of ASTM Standards.* Vol. 10.05, ASTM F84, ASTM, Philadelphia, 1988.

10. R. C. Eeast and M. J. Astle. *CRC Handbook of Chemistry and Physics.* 63rd ed., CRC Press, Boca Raton, FL, 1982–83, p. E-81.

11. F. M. Smits. "Measurement of Sheet Resistivities with the Four-Point Probe." *Bell System Technical Journal.* 711–718, May 1958.

12. S. J. Pearton. "Near-Surface Modifications Induced by Hydrogen in Semiconductors." *Proceedings.* Second International Conference on Solid State and Integrated Circuit Technology, Nov. 1989.

13. P. Kramer and L. J. Van Ruyven. "Space Charge Influence on Resistivity Measurements." *Solid State Electronics.* **20**, 1011–1019, 1977.

14. W. A. Keenan, W. H. Johnson, and A. A. Smith. "Advances in Sheet Resistance Measurements for Ion Implant Monitoring." *Solid State Tech.* June 1985.

Spreading Resistance Analysis (SRA) 24

ROGER BRENNAN and DAVID DICKEY

Spreading resistance analysis (SRA) is a method of dopant profiling that relies on the relationship between dopant concentration and electrical resistivity in semiconductors. In SRA, the resistivity profile is determined from resistance measurements made between two probes at a series of locations along a beveled surface of the sample. The method is used extensively on silicon, where it is generally an inexpensive way of obtaining a wealth of information. The accuracy for dopant concentrations is generally in the 10–30% range; and that for depth is typically better than 3%. The basic spatial resolution is on the order of microns, while depth resolution can approach 10 Å. SRA is sensitive to the full range of resistivity values found in both *n*- and *p*-type silicon. It is useful with a wide range of other materials, although the requirement for calibration seriously limits the accuracy in many cases.

Parameter measured	Electrical resistivity
Main usage	Determination of laterally resolved dopant concentrations for semiconductors, as a function of depth
Sample requirements	Surface of at least 20 × 20 µm area
Destructive	No, except for sample beveling
Detection limits	Carrier concentrations down to 10^{11} cm^{-3}
Quantitative	Yes, but calibration needed
Lateral resolution	A few microns
Depth resolution	~10 Å
Depth profiling	Yes, but requires beveling the sample
Instrument cost	Variable, but not less than $10,000
Size	Benchtop

Basic Principles

SRA determines dopant distribution by measuring the electrical resistivity of the silicon rather than counting atoms. Fortunately, resistivity is connected with dopant concentration in a semiconductor:

$$1/\rho = \mu q N \tag{1}$$

where ρ is the resistivity in $\Omega \cdot$cm; μ is the carrier mobility in cm^2/V$\cdot$s (V$\cdot$s = $\Omega \cdot$C, which can be obtained by substituting A = C/s into Ohm's law); N is the dopant concentration in cm^{-3}; and q is the charge of an electron (1.6021×10^{-19} C).

Measuring electrical resistivity therefore determines dopant concentrations. The spreading resistance probe has high spatial resolution and can detect variations in dopant concentration over depths as small as a few tens of angstroms. A test voltage is introduced into the specimen and the resistance is measured. Using the appropriate equations, one can extract values of resistivity from these simple measurements.

Methodology

Two probe tips, usually osmium, are mounted on the end of separate arms. Each arm pivots on a kinetic bearing system that virtually prevents lateral motion of the probe tip. The probe tips are positioned very closely together, less than 20 μm apart. The probe tips are lowered onto the sample as gently as possible, because of the high pressure of the probe tip—over a million pounds per square inch. The probe tip is hard enough to fracture the silicon, leaving "probe marks."

Typically, 5 mV is applied across the probes, and the spreading resistance is measured at a series of depths (see Figure 2). If the contact area of the probes is made small enough, the measured resistance is mostly from the current crowding in the immediate vicinity of the probe tip. This is determined by:

$$R_S = \rho/2a \tag{2}$$

where R_S is the measured spreading resistance in Ω; ρ is the local resistivity in $\Omega \cdot$cm; and a is the radius of the contact area of the probe in cm.

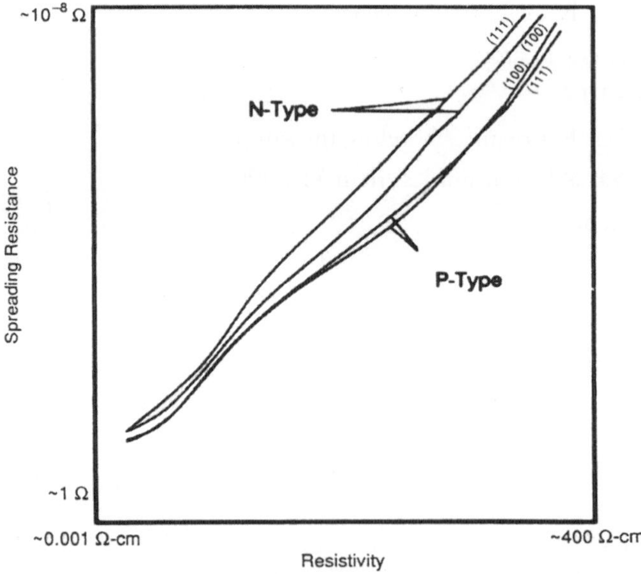

Figure 1   Calibration chart of spreading resistance versus resistivity.

This expression does not reflect most situations. Usually, calibration bulk samples with well-documented resistivity are measured and a is then calculated. The contact areas of the probes are seldom truly circular or the same size. When unconstrained, the resistivity is sensed in a sampling volume that approximates a hemisphere with radius a. Corrections are needed if there is an appreciable resistivity gradient through the sampling volume or if the sampling volume is distorted due to the proximity of a p–n junction, an insulator, or a region of much higher conductivity, such as a buried layer. Several data reduction procedures have been developed for these corrections.

An extensive calibration is usually required because of deviations from Equation 2. One calibration method would be to use 64 pieces with well-documented resistivities divided into four groups: $p<111>$, $p<100>$, $n<111>$, and $n<100>$. Each group has 16 samples ranging from 0.001 to 400 Ω·cm. In Figure 1 the calibration curves show departures from the straight line relationship of Equation 2. The n-type material produces higher spreading resistance readings than does the p-type of the same resistivity, and no material of a given resistivity produces higher spreading resistance readings than does $n<100>$. At low resistivities, the slope of the calibration curve becomes flatter as the resistivity decreases. At the high end of the calibration curve, the slopes are relatively constant and can be used to estimate resistivities greater than 400 Ω·cm.

Because the spreading resistance probe senses the resistivity in the sample direcdy under the probe tip, one can angle lap a silicon structure, probe down the beveled surface (see Figure 2), and obtain a resistivity-versus-depth profile. A plot of carrier concentration versus depth can be calculated using published mobility values.[1] Plots of carrier concentration versus depth are more useful because they tend to follow the doping profile. The spreading resistance probe does not directly sense the doping type. An additional test involves heating one of the probe tips and determining the polarity of the Seebeck voltage to find out conductivity type.

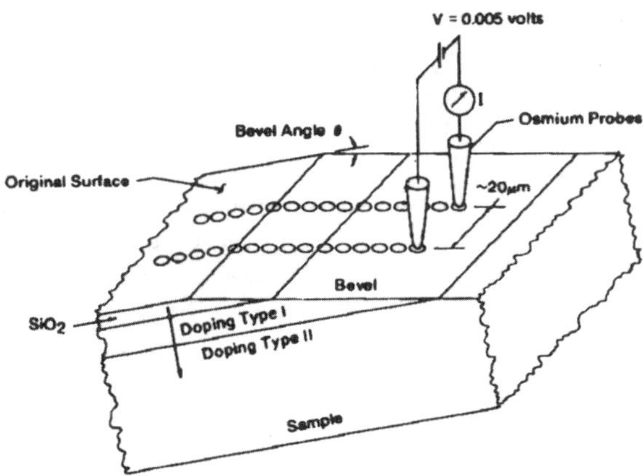

Figure 2   A spreading resistance measurement in a beveled sample.

Sample Preparation

Sample preparation is important. The sample must be beveled to the appropriate depth, and care must be used to avoid rounding the beveled edge, which would compromise the depth accuracy. Patterned samples should not be probed any closer to the diffusion mask edge than half the probe spacing. Also, the sample should not be beveled closer than half the probe spacing to the edge of the diffusion mask. Samples should be beveled immediately before probing. Long waiting periods, such as overnight, tend to produce noisier data. The lapping method should be totally mechanical to minimize nonstable surface charges.[2] The resultant bevel angle should be measured. Scratches on the beveled surface should be minimized because they change the probe contact area and thus compromise the calibration. Usually ion implants should be annealed prior to SRA.

Probing

Spreading resistance measurements are affected by noise, such as scatter in measured resistance on known homogenous material. The probe tips contact the silicon in different ways because the probes are lightly loaded and the contact area is very small. This causes differences in the measured resistance, but the noise can be reduced by increasing the load on the probe tip. Increasing the load increases the sampling volume and reduces the depth resolution. Typically, a probe tip loading of 2.5 g is used for shallow structures having junction depths less than 0.5 μm and a loading of perhaps 10 g for structures having junction depths 1 μm or deeper.

The probes must travel in a path relatively free of scratches and perpendicular to the bevel edge; the step size must be sufficient to prevent the probes from stepping into the previous probe marks. A well-prepared pair of probe tips should have low noise, a small sampling volume, minimum penetration, close spacing, and physical stability and compact size and should cause minimal damage to the silicon.

Data Reduction

The spreading resistance measurements are loaded directly into a computer containing the calibration data and the sampling volume correction algorithms. However, much human input is needed for data reduction of each profile run, for example, bevel edge, start point, doping types, points at which doping type changes, when to apply sampling volume correction, crystal orientation, step size, bevel angle measurement, adjustment for stray points, and measurement of probe separation. In addition, problems may arise with very shallow, very low dose surface layers due to geometrical effects and/or surface charges.[3]

When differences in resistivity are relatively small, the relatively high noise of the spreading resistance measurement makes data reduction uncertain. Data reduction is difficult at low resistivities due to the flatness of the calibration curve in this region. At concentrations above 10^{19} cm^{-3}, SIMS can be more accurate than SRA, but SIMS senses atomic concentration and SRA senses carrier concentration. There can be a large difference between these measurements at low resistivities; however, sometimes a combination of SRA and SIMS is needed.

Applications

Typical applications of SRA include predepositions and drive-ins, epi thickness and profiles, full bipolar transistor structures, starting material, buried layer-up diffusion, poly resistivity and thickness, autodoping, and ion implant doses and depths. SRA is useful in checking the quality of incoming material, developing and documenting a new process, modifying a process and confirming the results, checking the quality of production processes, and finding out why the wafer sort yield went up or down.

Dedicated Spreading Resistance Test Patterns

To do SRA, one needs to identify a suitably large area containing the desired structures. The area needed—one of the smallest areas required by any profiling technique—can usually be found on the chip, but this is often less than optimal. As a long-term solution, dedicated spreading resistance test structures are suggested (see Figure 3). Sooner or later, a new test pattern will be laid out anyway, and often these features can be included with negligible time and expense. The spreading resistance test patterns offer minimal analysis cost, complete characterization of the wafer fab process, optimum resolution, and clear, straightforward identification of all possible doping structures. By selectively exposing the silicon at the appropriate masking steps, one can produce rectangles that contain every structure of possible interest.

Comparison with Other Techniques

SRA Versus SIMS

To characterize, for example, the deep-diffused power transistor referred to in Figure 4 requires seven decades of dynamic range; SIMS is considerably more limited (five decades). The power device needed to be profiled down to 120 μm; this is too deep for SIMS to depth profile economically. SRA detected a high resistivity region from about 40 to 60 μm; few, if any, other techniques can detect this.

SRA Versus MOS C–V and SIMS

Analyzing the blocking ability of a material (such as photoresist, oxide, or nitride) to prevent dopant from getting into the silicon requires checking four possibilities: (1) no dopant penetration into the silicon, (2) some dopant penetration, (3) more dopant penetration, and (4) considerable dopant penetration. SRA is the only way to catch all these possibilities in one pass. MOS C–V would characterize (1) and (2) very well, but would have trouble with (3) and serious difficulties with (4); SIMS would be helpful on (4) only.

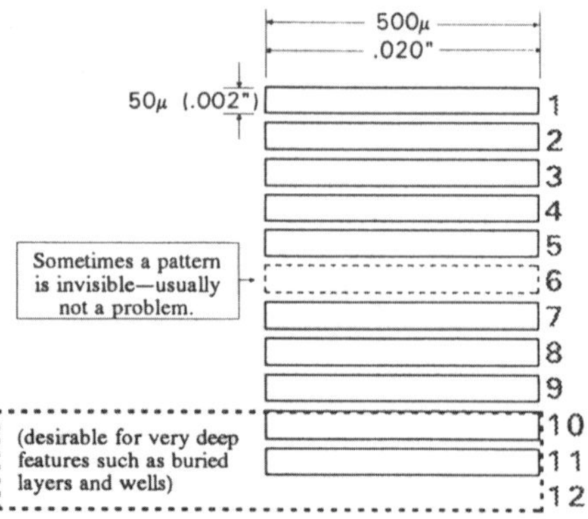

NOT ACCEPTABLE

NOTES

1. If there is a very shallow emitter/base and a very deep buried layer a $100\mu \times 1000\mu$ structure is suggested.
2. In general, longer is always better—say up to 1500μ.
3. Width can be a problem as the number of patterns increases. Please reduce the space between patterns to what is essential for lateral diffusion effects.
4. Label the structures at the first possible masking step.
5. Orient the long dimension of the pattern parallel to the major flat.. This is strongly recommended for epi on <111> substrates.
6. Make one common buried layer when possible. (Also wells.)
7. In relatively high beta (narrow base) bipolar devices the base width should be characterized with and without the buried layer underneath it.
8. Top side passivation is an annoyance for us (causes scratches and slows down beveling.) Please *consider* etching the entire region at pad mask. Perhaps some etching tests should be made first.
9. Sometimes allowances for "invisible" patterns, large epi shifts, and huge silicon step heights need to be made.
10. One or more back-up sets would be appreciated.
11. After dice separation, you may wish to keep the test dice as part of the permanent record for that lot of wafers.
12. If all dice must be product dice and you are exposing with a stepper, you could place these structures two high in the scribe lines. It is better than nothing.

This list of notes is surely incomplete. Although SRA was developed in the early sixties, the technique continues to evolve (i.e. there are still more subtleties to identify, resolve and document).

Figure 3   Illustration of dedicated spreading resistance test patterns.

Limitations

Only the net *n*- or *p*-type concentration of the electrically active dopants are sensed by SRA. SRA is destructive and highly technique-dependent and can have substantial uncertainty in resistivity measurements. C–V analysis may be more accurate for

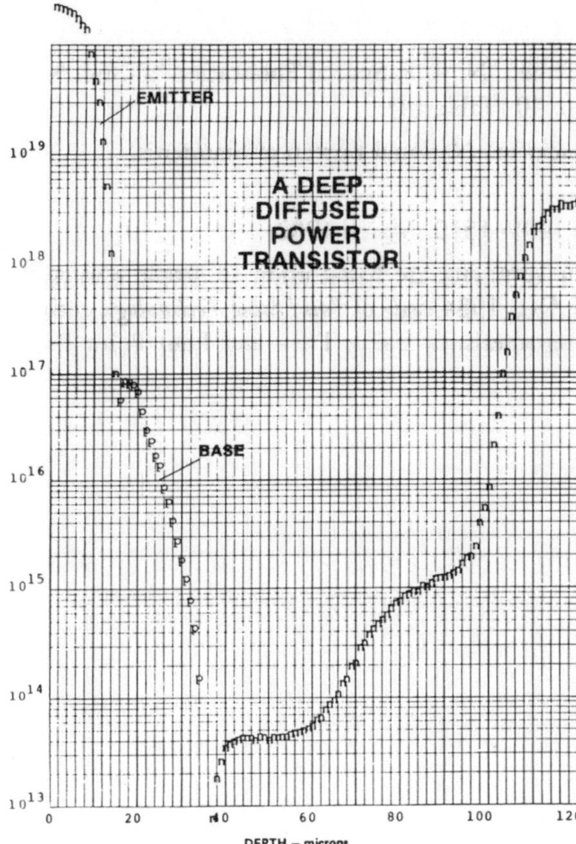

Figure 4   A deep-diffused power transistor

channel regions in MOS devices, and SIMS may be more accurate for concentrations greater than 10^{19} cm^{-3}. In addition, problems have been experienced with very shallow, very low dose surface layers due to geometrical effects and/or surface charges. Equipment maintenance and calibration are relatively involved; without professional dedication, many serious mistakes can be made.

Conclusions

The dopant distribution controls almost all electrical properties of semiconductor devices. To produce a semiconductor device with the desired electrical properties, the dopant distribution must be known. SRA monitors dopant distribution over more than eight decades; no other single analytical technique can match this range. SRA also provides general purpose, and relatively inexpensive, depth profiling with high spatial resolution and excellent depth accuracy and with unmatched sensitivity at low doping levels.

References

1 Thurber, Mattis, Liu, and Filliben. N. B. S. Special Publication 400–64, May 1981, p. 34 (Table 10) and p. 40 (Table 14).

2 J. Ehrstein. Private communication. National Bureau of Standards, Gaithersburg, MD.

3 S. M. Hu. *J. Appl. Phys.* **53**, 1499–1510, March 1982.

Several sections of this technique summary are based upon an article written by R. Brennan and D. Dickey in the Dec. 1984 issue of *Solid State Technology*, entitled "Determination of Diffusion Characteristics Using Two- and Four-Point Probe Measurements."

In Static Secondary Ion Mass Spectrometry (Static SIMS), a primary beam of ions (typically 1 to 10 keV) penetrates into a solid surface. The energy transferred to the lattice causes atoms, or clusters of atoms, to eventually be ejected from the surface region (from the top 1 to 3 atomic layers) as either ions or neutrals. These are mass analyzed in a mass spectrometer (either quadrupole or time-of-flight, TOF), allowing their identification. The difference, compared to dynamic SIMS (see the Dynamic SIMS summary), is that the dose of incidence primary ions is kept low (less than 5 times 1012 atoms/cm^2) so that, during analysis, the chemical integrity of the surface is maintained. Essentially every incoming ion strikes a fresh area not previously impacted. The objective is to deduce the chemistry of the outer layers by interpretation of the fragment ion pattern observed (cf. classical organic mass spectrometry).

The mass range accessible is up to 1000 amu with a quadrupole and 10,000 amu with TOF. TOF also has high enough mass resolution so that there is no ambiguity in atomic, or cluster identification (because atom isotopes have fractional masses, allowing any combination of masses to be distinguished with high enough mass resolution). The spatial resolution is controlled by the focus and energy of the ion beam, and is typically a few tens of μm, but sub-μm can be achieved in some cases.

The major use is in identification of the molecular species present at surfaces, particularly for polymers and organics. This may be for components of bulk materials (e.g., in polymer blends), or for monitoring changes caused by surface treatment, or for detection of contamination. Trace levels can sometimes be reached, but quantification is difficult and requires standards. With the high sensitivity and resolution of the TOF, the technique can also be used for detecting and quantifying trace heavy metals at surfaces.

Surface roughness can be measured by a variety of scanning profilometer methods, which monitor individual features, or by methods which give averaged information. For the profilometers there is a wide variety of methods possible: mechanical, optical, scanning electron microscope (see the SEM summary), and the scanning force microscope and scanning tunneling microscope (see the STM and SFM summaries). These instruments have a large range of capability in terms of depth resolution, maximum depths measurable, and lateral resolution (roughness frequency range). The mechanical profilometers obviously contact the surface, and may damage it. The depth resolution can be down to 5 nm, and the lateral resolution 100 nm (depends on tip radius), for the most sophisticated (and expensive). Optical profilometers can measure down to 0.1 nm depth and 300 nm lateral resolution (depends on wavelength).

SEM is useful for observing very short range roughness (i.e., sharp spikes or dips) but ineffective for long range small amplitude roughness. SFM has depth resolution down to 0.01 nm and a normal lateral resolution of about 0.5 nm, which is more than needed in most practical applications. STM can do an order of magnitude better, but only for conductors under ideal conditions in UHV environments. It is basically a research tool.

Optical scatterometry (see the summary for this technique) gives averaged, not individual, information. The depth resolution achievable is 0.1 nm RMS, and long and short range roughness can be separated (i.e., measurement provides information as a function of roughness frequency range).

Total Reflection X-Ray Fluorescence (TXRF) is a technique specially developed to detect trace contamination elements at wafer surfaces. It detects and quantifies their presence from their characteristic X-Ray Fluorescence, as in XRF (see XRF summary). The difference from normal XRF is that the X-ray beam strikes the surface at a very grazing angle (a few mrad) below the critical angle for total reflection, so that the beam never penetrates far into the material. The depth probed is only a few nm and depends on the surface roughness. In order to work, the method obviously requires a very flat surface, such as a Si wafer, and the grazing angle results in the X-ray beam intersecting a cm or more length of surface.

The main use is rapid multi-element analysis with very good trace detection limits for heavy metals (down to 109 atoms/cm^2, depending on element). Quantification is possible using standards. By varying the angle of incidence below and just up to and above the critical angle, some limited depth resolution capability is possible, including distinguishing atomic layer type contamination from 3D contamination, such as particles.

In Transmission Electron Microscopy (TEM), a very high energy monoenergetic electron beam (100 to 400 keV) passes through a thin specimen (less than 1000 nm) of diameter less than 3 mm (necessary to fit in the electron optics column). A series of post specimen lenses transmits the emerging electrons, with spatial magnification up to 1,000,000, to a detector (fluorescent screen or video camera) viewed in real time.

Any regions of the sample under the incident beam (usually a few μm diameter) exhibiting crystallinity, will diffract electrons away from the central spot, forming a diffraction pattern observable at the back focal plane of the objective lens. Like X-ray diffraction, this can provide identification of crystalline phase, orientation, and lattice parameters. In micro-diffraction, the incident beam is focused down to sub-micron areas, but this focusing degrades the diffraction pattern.

The above mode of operation is termed *Diffraction Mode*. Another mode is *Imaging Mode*. In this mode the reconstructed, highly magnified image (up to 1,000,000) is observed at the image plane of the objective lens. By placing an aperture in the diffraction plane, either the electrons in the undeflected beam (the center spot) can be imaged (Bright Field Mode), or those in a deflected beam (the diffracted or scattered electrons) can be imaged (Dark Field Mode). An amorphous sample of uniform thickness can undergo only Z contrast scattering (higher Z, more scattering), so regions of higher Z show up as dark areas on a bright background in Bright Field Mode. Any crystalline region will also show up as a dark area. In dark field, regions of crystallinity, or higher Z, will show up as bright spots on a dark background.

Yet another mode of use is High Resolution TEM (HRTEM), in which atom positions can be established by collecting electrons from both the undeflected and diffracted beams and comparing the observed phase interference patters to a simulation.

TEM, with its many modes, and often involving ancillary materials analysis capabilities such as EDS (see EDS summary), is the mainstay of material science and analysis of small volume (areas and thickness). A fully equipped TEM laboratory will have several microscopes with differing capabilities, plus all the necessary sample preparation techniques. See also Scanning TEM (STEM), where the incident beam is focused down to almost atomic dimensions and scanned across the sample.

In ellipsometry, a collimated polarized single wavelength light beam (UV, visible, or IR) is directed at the material under study, at an oblique angle, and the relative light intensity of the reflected beam is measured as a function of its polarization angle. In Variable-Angle Spectroscopic Ellipsometry (VASE), these measurements are made as a function of varying both the wavelength and the angle of incidence. For a thin film with a surface parallel to the substrate (i.e., a constant thickness) and inhomogeneities of less than about 1/10 of the wavelength being used, Maxwell's equations plus the Fresnel equations for calculating reflection coefficients at interfaces can be used to model the experimental data. The variable parameters in the modeling are layer thickness and the optical constants for the material in the layer. Surface and interface roughness and/or mixing and interface reaction can, in principle, be handled by treating the interfaces as additional layers with their own optical properties and thickness. The parameter values in the model are itterated until a best fit with the data is obtained. Though data is easy to obtain, this model-dependent aspect of ellipsometry analysis, with many floating parameters, can make it a difficult technique to use correctly (it can never give a unique fit), unless some of the parameters (such as the optical constant) are already well known, and can therefore be constrained.

VASE, and also single wavelength ellipsometry (less reliable) are extensively used in the wafer processing and other thin film industries, primarily to monitor thickness of single or bi-layers of dielectrics, optical coatings, semi-conductors, and magneto-optic and magnetic disk material. Depending on the optical constants, film thickness from 1 to a few 100 nm can be handled (or even microns for transparent material). Because the instrumentation is operated in air and the light can penetrate liquids, ellipsometry can also be used for in situ electrochemistry and biological interfaces.

In X-Ray Diffraction (XRD), a collimated beam of mono-chromatic X rays between 0.5 Å and 2 Å wavelength strike a sample and are diffracted by crystal planes present. Bragg's law,

$$\lambda = 2d\,\sin\theta$$

relates the spacing between planes, d, to the diffraction angle, 2θ, which is scanned to pick up diffraction from the different crystal planes present. The azimuthal orientation of the different beams also reveals the crystalline orientation. If the material is poly-crystalline, then diffraction rings, instead of a spot pattern, are formed (powder diffraction). Distortions or broadenings of the diffraction beams carry information on crystal strain and grain size.

Because X rays penetrate deeply (strongly Z dependent; at $\lambda = 1.54$ Å the absorption length is 1 mm for carbon, 66 µm for Si, and 4 µm for iron), XRD is intrinsically a bulk technique. Typically, however, large areas are used (several mm diameter), and there is sufficient intensity and spectral resolution using modern standard equipment to study films of thickness down to the 100's nm range for high Z material. If specialized instrumentation and geometries are used (Grazing Angle XRD, Double Crystal Diffraction XRD, synchrotron radiation source), sensitivities down to 10's nm are easily achieved, and mono-layer sensitivity is possible. Small area capabilities also exist (micro-beam XRD), but then thicker samples are required to compensate for intensity loss of the smaller areas.

The major uses of XRD are identification of crystalline phases, determination of strain, crystalline orientation and size, epitaxial relationship, and the accurate determination of atomic positions (better then in electron diffraction). Because of the strong Z dependence of X-ray scattering, light elements are difficult to deal with, particularly in the presence of heavy elements.

In X-Ray Fluorescence (XRF), the mono-energetic X-ray beam from a vacuum X-ray tube (W, Mo, Cr, and other targets are used to provide a range of energies) irradiates the sample, in air, causing excitations of core electrons of the atoms present. As these excited atoms decay back to their electronic ground states, light is emitted in the X-ray region. The specific wavelengths emitted are characteristic of the energy levels involved, and therefore of the atoms concerned. The X-ray fluorescent wavelengths are determined using a crystal diffractometer.

The depth probed is Z dependent, but, since X rays are involved both in the excitation and the emission, it is many μm, making XRF a bulk technique. A specialized grazing incidence (total reflection) adaptation exists, however, for analyzing material at the surfaces of semi-conductor wafers (see TXRF summary). Also, there is sufficient total intensity such that thin films (μm level or lower, depending on instrumentation and Z) can be analyzed provided there are no interfering elements in the substrate. Standard instruments have no significant spatial resolution (few mm), but micro-beam systems down to 10 μm do exist. The technique is applicable to all elements except H, He and Li. Energy dispersive X-ray analysis (see EDS summary) is a closely related technique.

The major uses of classical XRF, in air, are the identification of elements and the determination of composition for bulk materials. For thin films intensity/composition/thickness equations are used to determine the composition and thickness of individual layers in single or multi-layered stacks, such as used in the disk drive and semi-conductor industry.

In X-Ray Photoelectron Spectroscopy (XPS), also known as Electron Spectroscopy for Chemical Analysis (ESCA), mono-energetic soft X rays (usually Al K alpha at 1489 eV or Mg K alpha at 1256 eV) bombard a solid sample in ultra high vacuum. One of the interaction processes occurring is the ejection of photoelectrons, according to the Einstein Photoelectric Law:

$$KE = h\nu - BE$$

where $h\nu$ is the incident X-ray energy, KE is the kinetic energy of the photoelectron, and BE is the binding energy of the electron in the particular energy level of the atom concerned. Measurements of the KE's of the ejected photoelectrons directly determines the BE's, which, for core levels of atomic orbitals, are characteristic of the atoms concerned and therefore identify their presence. On a finer energy scale, small "chemical shifts" in the BE's provide chemical state identification (such as the oxidation state for a metal). The technique is applicable to all elements, except H and He, since they do not have characteristic atomic core levels.

For a solid, XPS probes 2 to 20 atomic layers deep, depending of the KE of the ejected electron, the angle w.r.t., the solid surface of detection, and the material. XPS is therefore a true surface technique. If measurement is made as a function of angle, depth distribution information is available over the depth probed. XPS is also used with sputter depth profiling to go beyond these depths. Modern laboratory XPS instruments can provide *practical* lateral resolution capability down to the 10 to 20 μm range. Specialized systems can go lower, but acquisition time becomes very long.

The particular strengths of XPS are semi-quantitative elemental analysis at surfaces without standards, quantitative analysis with standards, and chemical state determination for materials as diverse as biological to metallurgical.